INNOVATION ELEGANCE

Transcending Agile with Ruthlessness and Grace

ROBERT F. SNYDER

"Like a splash of cold water, *Innovation Elegance* demands your attention from the very first chapter. It is refreshingly honest about why current approaches to innovation fail and how the Elegance methodology, its concepts, principles, and even choice of words (Five Verbs) help change leaders not only overcome barriers to success but develop high-trust, high-performing teams. The author's insightful perspectives, analogies, and engaging writing style make this a must-read for project, change, and innovation teams."

– Jessica Crow, change management and organizational effectiveness expert, speaker, and consultant

"Let's face it, project success rates are not good. *Innovation Elegance* gives insights into the causes and solutions. As we all know, effective communication is critical for project success and the idea of an agreement factory is a great way to highlight the power of ensuring agreement before starting a critical project. It also gives us a way to ensure that we stay in agreement throughout all the rough spots we'll find along the way. I think about things differently now which is the hallmark of a book worth reading."

– John Fisher, CIO, entrepreneur, and board director

"*Innovation Elegance* brilliantly explores the connection between the arts and innovation, revealing a world where discipline, empathy, and collaboration intertwine to create something truly remarkable."

– Nazia Raoof, transformation leader

"Author Robert Snyder has a truly innovative idea: a methodology for innovation *teams* rather than a methodology for building a *whatever-it-is*. Bringing the team centre stage promotes people and interactions, which is where the magic of innovation actually happens."

– Allan Kelly, Agile and OKRs consultant, keynote speaker, and author

"*Innovation Elegance* masterfully cuts through the noise, offering not just a roadmap for innovators grappling with cultural pain points, but an entire GPS system for navigating the complex terrain of modern innovation. The performing arts analogy, as well as tying key practitioner themes together in new ways, is original, meaningful, and thought provoking. It's a fresh and valuable take on where the rubber meets the road for those committed to leading organizations with both discipline and empathy."

– Marian Cook, transformation leader and educator, public and private sector

"*Innovation Elegance* is a philosophy and methodology that aims to return sanity and rigour into a world turned Agile, marrying discipline and empathy into an exceptionally powerful and effective combination. This book was a delight to read. There are so many pearls of wisdom, uncommon insights and valuable experience packed in these pages, I highlighted something memorable on almost every single page."

– Bard Papegaaij, leadership philosopher, author, and coach

"*Innovation Elegance* captures the very essence of what it takes to be an innovation success–the combination of a people-centric, customer-centric, and employee-centric methodology. Innovation does not come from a software program but rather from people. Creating a culture of collaboration maximizes people's potential to innovate."

—Tom Kuczmarski, President of Kuczmarski Innovation and co-founder Chicago Innovation.

Dedication

My first dedication of this book is to my brother Paul and my good friend Kevin, who passed away during early drafts of this book. Their passing encouraged me to get this material on paper sooner rather than later.

My second dedication of this book is to empathetic Change Agents of the 21st Century. I hope this book transfers all the confidence I have in this material to you and your team as you pursue healthy change. I hope you make it your own, improve upon it, and pay it forward.

**Upcoming Books by Robert F. Snyder
in the Innovation Elegance Series**

Innovation Portfolio:
Five Verbs Shape Your Team's Legacy

Elegant Leadership:
Distinguishing The Good, Bad, and False

Acknowledgements

Thank you to Kinga, Vinay, Toby, Dan, Michele, Bianca, Lisa, Brian, Albert, and Sylvia. Their early feedback and encouragement eased the early work.

Special thank you to Kathy and our thoughtful feedback discussions at Starved Rock State Park.

Special thank you to my developmental editor, Dr. D. Olson Pook, who walked a couple of tightropes with me to reorganize the material so readers could make sense of it.

Kathy, Olson, and the others made this instrument the best it could be and set the table for sequels (books, conversations large and small) to set new frontiers in healthy change.

Contents

Preface

Nothing will change until the status quo is more painful than the transition.

~ Laurence Peter (1919-1990), Professor of Education,
University of Southern California

To me, this book was inevitable. If I didn't write it, someone else would have. To use a hockey analogy, this book shows where the puck is going for innovation methodology. But since no one else assembled these ideas in book form, here we are.

The target audience for this book is innovation professionals whose cultural status quo causes pain. That pain takes the form of poor culture, poor methodology, poor management, or poor leadership. Project managers, change managers, and business analysts who feel good about their culture will shrug at the idea of a new methodology. But for many innovators and project professionals, the status quo is sufficiently painful for them to be early adopters of this methodology.

When I first put pen to paper, I only intended to write a book to help reduce team frustration and high project failure rates. I wanted to provide rigorous tools for both left-brain (logical, analytical, orderly) and right-brain (intuitive, imaginative) lines of thinking. A long journey of revisions led me to propose a methodology to evangelize innovation literacy, improve employee experience, and infuse the culture traits of the performing arts.

Feedback on these early concepts led me to conclude that directly challenging the status quo was the most authentic approach. A modest set of tools and references to the arts evolved into the audacious goal of firmly confronting Waterfall, Agile, and their software-centricity.

I didn't want to directly confront Agile so much as casually stroll past it. However, since Waterfall has enough critics, I am positioning this methodology as transcending Agile. This book presents a people-centric methodology with an unusual pairing of ruthless discipline and empathy, a pairing that is full of grace to achieve healthy change.

The seeds for this methodology first appeared to me in 2005. During the day, I worked as a typical IT (information technology) project manager. In the evenings, I was earning my MBA. An early favorite course of mine was Basic Operations. The group assignments and case studies covered organizations such as bicycle assembly shops and cranberry processing plants. The language included terms such as speed, quality, inventory, variability, and waste. The program aimed to teach the fundamentals of a factory: how it promotes motion, discipline, a sustainable pace, and tangible output.

Similarly, a well-functioning factory avoids motionlessness, fragmented expectations, 'firefighting,' and chaos. I concluded, "Well, *every* project team is a kind of factory—an agreement factory."[1] Effective teams agree on things such as assignments, the next project, a new product or process, and testing and training activities.

In contrast, dysfunctional teams can be called *disagreement factories*. It's easy to understand how a culture of interruptions, mistrust, and poor transparency hurts factory speed. A culture with heavy favoritism, personality conflicts, and a lack of purpose hurts factory quality. Poor focus, counter-

---

1   'Agreement' here has a positive connotation, not to be confused with 'groupthink' or mindless consensus. If you prefer, think of it as being in 'alignment.' In contrast, a dysfunctional team would then be a 'misalignment factory.'

productive documentation, and high employee turnover are forms of waste in an innovation factory.

With this conclusion, I ran my project teams as *agreement factories*. This metaphor of a factory was the first ingredient to shape the methodology.

The second ingredient involved the form these agreements take. Over the next few years, as Agile methodology became common, then the norm, documentation fell out of favor. Team agreements defaulted to the form of meetings and email. Information sharing and collaboration became more laborious. Communication traffic jams became the norm.

Of course, teams need more than only verbal agreements. However, agreements that reside solely in email are also insufficient, since too often, the right people are not writing or receiving the email. Sometimes, an email propagates a cover-your-backside (or CYA) culture, because Person A is typing *at* Persons B, C, and D.

I concluded that teams cannot just talk and email their way to success. These agreements needed to form and reside in a different format. They needed to be memorialized in a document, an artifact, a deliverable—something the team had obviously collaborated on and could physically *print* if needed. Structured documents placed that collaboration on a pedestal for everyone to see. When my team formalized an approach to documentation, we minimized blind spots, ambiguity, and rework. Documentation governed in a standardized way was less at the mercy of someone's email backlog and last-minute meeting conflicts.

Although the number of documents might be significant, each document has a manageable size. Documents also easily accommodate additional and previously forgotten stakeholders. These documents are not rigid; they set expectations for the moment and so comprise an *expectation-setting factory*. In contrast to the fleeting value of a meeting or an email, agreements formed via a framework are valuable for a long time—in other words, they become *assets*. In addition to leading an *agreement factory*, I felt I was leading an *asset factory.* In building and maintaining dozens of assets, my teams were building asset portfolios.[2]

An asset portfolio is a concept that promotes modest risk, high reward, and long-term health. An asset portfolio avoids single points of failure. A culture of documentation promotes simplicity, transparency, and a sense of accomplishment. Project-independent documentation encourages prioritization, humility, and listening, while project-specific documentation establishes interdependencies, pace, and accountability.

Yet on their own, the two themes of an agreement factory and an asset portfolio felt dry and sterile. They lacked charm and inspiration. The seriousness of these left-brained metaphors begged to be balanced with an empathetic, right-brained approach.

In 2012 I started taking Latin dance lessons. I had ample time and passion and learned quickly. I joined teams, performed, and even began teaching classes to beginners. Most importantly, I saw that everything I learned on the dance floor could be applied off the dance floor (i.e., to innovation teamwork).

Although my work teams weren't literally dance teams, we needed clarity about who was leading and who was following. We were at our best when we followed a rhythm, remained elastic (not rigid or limp), and stayed aware of other people on the dance floor.

It was easy to see the value of cross-pollination with other artistic formats such as music and theater. Although my teams weren't literally symphonies, we needed to listen, balance, and achieve harmony. Although we weren't literally theater companies, we needed a great script, actors to fulfill their roles, and audience (customer) centricity. The arts possessed countless culture traits I wanted my project teams to aspire to and emulate.

And I believed there was value in stretching the metaphor even further. Although my teams were not literally improvisation teams, we needed skills to think on our feet, freedom to make mistakes, and commitment to each

---

2   The methodology chooses the term asset instead of investment. Investments are bought and sold; their value rises and falls. A project team doesn't sell its assets, and team assets retain their value.

other's success. Although my teams were not raising children, we needed to emulate the art of parenting, such as investing in beginners, building self-esteem, and teaching self-sufficiency.

And finally, given that people fight, we needed rules for conflict. We needed to harness task conflict and neutralize personality conflict. We needed to disagree without demonizing. We needed to navigate organizational politics, opponents, and bad actors *with empathy*. We needed the grace and the ruthlessness of martial arts.

As a third ingredient, these empathetic arts contain numerous culture traits valuable to teamwork and collaboration. In cross-pollinating these activities, I believed everyone could find something to improve their contribution, value, and morale.

The cross-pollination you will see emerge in the methodology is already happening in countless worldwide forums. These three themes pop up as one-liners, punchlines, and motivational quotes. You hear factory references such as a "surplus of this," "shortage of that," and "hurry up and wait." You hear clichés about the arts such as "singing from the same hymnal," "it's a delicate dance with this client," and "let's improvise."

These great quips hint at where the puck is going, but they tease. They only scratch the surface of the potential in cross-pollination, constrained as we are by sixty-minute virtual meetings and twenty-first century attention spans. My proposed methodology blends these morsels of wisdom, shamelessly bathing itself in the three metaphors. This cross-pollination shapes a team culture of discipline and empathy. And as this methodology took shape, the title that continuously felt right was Elegance.

The Elegance methodology shapes culture by blending these three metaphors into practice: culture disguised as a factory, culture disguised as an empathetic art, and culture disguised as asset-generating documentation. Teams that apply the tools behind these metaphors shape a culture of discipline and empathy.

An effective innovation methodology needs all these elements working in harmony. Existing methodologies provide only a half-education for team discipline. In comparison, the factory and asset portfolio metaphors of the Elegance methodology spell out education for team discipline to the point of ruthlessness.

Existing methodologies also ignore empathy and elegance. Content related to the arts and leadership educate about beauty, compassion, and grace. Ruthlessness *and* grace? Yes, navigating our complex world requires both. The best leaders are aggressive with process and gentle with people. Leaders can be demanding about documentation but exercise a soft touch as the team performs the work. Good leaders know when to bang the table and when to set the table. These themes are the necessary combination to achieve Innovation Elegance.

Over the past few years, frustration in the innovation space has risen. I propose that integrating these topics—the factory, the asset portfolio, the empathetic arts, and effective leadership—will make a difference. Within my own team, I educated colleagues on these themes, and we gradually adopted the framework and tools. I saw conclusive effects and was encouraged by the positive feedback.

While blog posts and LinkedIn articles are great ways to share small slices of the methodology or a single tool, formalizing the Elegance methodology in book form felt best. I hope this format enables a comprehensive understanding of how these valuable themes reinforce each other and turn frustration into fulfillment.

INTRODUCTION: A PEOPLE-CENTRIC METHODOLOGY

You don't build a business.
You build people and then people build the business.

~ Zig Ziglar (1926-2012), American author, salesman,
and motivational speaker

This book centers on a single, controversial claim. It's likely to ruffle feathers.

What is it? It's the assertion that, more often than not, your organization's innovation methodology is setting you up for failure.

That's right: Agile sets you up for failure.

"Not so!" you say. Your organization is profitable, and you've had success. You've completed many projects using the methodology and, while nothing's perfect, you're sure your success rate would be lower without it.

But I would argue that while you succeeded in implementing change, it was in spite of the methodology you used, not because of it.

You might be a for-profit company executing a 'mostly' Agile methodology. You might be a nonprofit with a few humble databases and with nothing formalized. You might be in government, making the leap from Waterfall to Agile. Whatever the case, you are running a software-centric methodology to implement change in your organization because that's what everyone else is doing.

That is entirely understandable. Management training, leadership books, and professional certifications for these software-centric methodologies abound. But over twenty years into their use, project failure rates (sometimes cited at 50 to 70 percent) remain unacceptably high.[3]

The world has used these software-centric methodologies for two decades. Difficulties still abound. It's time to reconsider your devotion to your current methodology and fix this problem with 21st century thinking and values.

---

3   Definitions and statistics about project failure vary significantly; dozens of information sources exist, and the statistics vary year-to-year, so I do not emphasize specific sources here. Some of the higher-profile sources for this number include Project Management Institute (PMI), CIO magazine, and *Harvard Business Review*.

Defining the Problem

The first step in solving a problem is to recognize that it does exist.

~ Zig Ziglar

What makes innovation difficult is not software. What makes innovation difficult is people.

If you're reading this book, your team is likely a competent, experienced, well-trained, and well-intentioned crew. Despite the considerable skill set they bring to the table, change and innovation in your organization are more difficult than they should be. Despite great individuality, the problem is how we treat each other, govern, and collaborate. Your methodology must confront this reality, not yield to a decades-old emphasis on software.

The professions around innovation have yet to supply a people-centric methodology. This book does. Based on the insights of a thirty-year career focused on innovation, the Elegance methodology orchestrates a rhythm of healthy change. It steers innovation professionals towards customer value, a rewarding employee experience, and outstanding teamwork. It empowers innovation professionals to convert a dysfunctional environment into a sane, sustainable, stimulating culture.

Innovation Elegance is people-centric in three ways. First, it is customer centric. For many companies, customer centricity is already a buzzword, a priority—even 'in their DNA.' But existing methodologies make it easy for you to *say* you are customer-centric while not *being* very customer-centric. The Elegance methodology turns the screws on that discrepancy. It aims to give customers great value, wonderful experiences, and memorable stories.

Second, Innovation Elegance is employee centric. Innovation projects are often painful for employees, but software isn't the root cause of that pain. The Elegance methodology cultivates safety, confidence, focus, and enthusiasm to optimize the employee experience.

Third, Innovation Elegance anticipates people-based problems and instills strong governance for teamwork and collaboration. A stakeholder might want a project to fail because they fear change. A hyper-competitive employee might undermine their colleagues. Or an unskilled manager, poor leader, or lousy boss might create a toxic environment. The Elegance methodology contains tools to neutralize and survive these contrarians.

A Brief History of Innovation Methodologies

Dear Past, Thank you for the lessons.
~ Unknown

"Set you up for failure" is a provocative statement. A gentler version is, "Your methodology limits you." Let's understand these limitations by first recalling the history of innovation methodologies.

Between approximately 1980 and 2000, the predominant methodology was Waterfall. Waterfall's reputation was—and is—an emphasis on documentation, large projects with spread-out completion dates, and rigidity against changes mid-project. Companies were satisfied with Waterfall so long as market competition and the pace of change were mild. Companies accepted high-risk, lumpy, big-bang go-live events many months apart.

Near the end of that window, the internet rose in relevance, and the demand for building websites skyrocketed. Established companies and start-ups alike frantically built their own shallow brochureware on the World Wide Web.

Changing sophisticated business logic that evolved over the years could allow for months of work. Waterfall was a good fit for that kind of work. But changing webpages to provide a simplistic tour of the business in only a few days of work led to Waterfall falling out of favor. Businesses wanted a new methodology that would keep up with the desired pace of change. That new methodology was named Agile.

Agile was formalized by the Agile Manifesto in 2001 by software developers. Like Waterfall, Agile emphasized software. But unlike Waterfall, which prioritized documentation before the phase of work typically called Build or Code, Agile de-emphasized documentation which, during the early web craze, was seen as less valuable. The third value in the Agile Manifesto illustrates this, favoring "working software over comprehensive documentation." The manifesto's seventh principle declares, "Working software is the primary measure of success."

The Agile methodology has a singular responsibility—to deliver code (in the form of new features) approximately every two weeks. Again, it's all there in black and white in the Agile Manifesto:

- The first principle: "Customer satisfaction by early and continuous delivery of valuable software."

- The third principle: "Deliver working software frequently (weeks rather than months)."

- The eighth principle: "Sustainable development, able to maintain a constant pace."

In the first twenty years of this century, the sophistication of websites and smartphone apps has skyrocketed. Agile has evolved and tried to handle this sophistication through spinoffs like Disciplined Agile, Scaled Agile, DevOps (Development and Operations), and DevSecOps (Development and Operations, with Information Security added in for good measure).

At the time of this book, the most common term organizations use to label their methodology is Hybrid. Hybrid vaguely combines Waterfall and Agile, sometimes with cute names like Wagile and Agi-fall.

Companies adjust and compromise where Agile is a poor fit, but Hybrid is still software-centric: it has its limitations and sets you up for failure.

The Limitations of Agile and Hybrids

When solving problems, dig at the roots instead of just hacking at the leaves.
~ Anthony D'Angelo (b. 1955), American business owner,
author, and motivational speaker

The essence of the Elegance methodology is people governing people with discipline and empathy to sustain a working team. Agile and Hybrid methodologies are limited in this way because they neglect people and emphasize working software. Here are five ways Agile and Hybrid neglect people and emphasize the wrong things.

#1: Frequency Profile Is Not Cost Profile

Foremost is the insinuation that technical agility equals financial agility. It does not. Through the lens of a technologist, agility pertains to frequency: delivering code every two weeks. Through the lens of an economist, agility pursues minimal marginal cost in the long run by accepting a higher upfront cost in the short run.

The Agile methodology governs short (usually two week) sprints for projects and embraces short-term decision making. But Agile's dedication to frequency has a dark side when impulsiveness, interruptions, and chaos try to disguise themselves as agility. Organizations that abuse the label 'agility' and try to squeeze more value into a frequency-based methodology cause churn, fatigue, and burnout—not better financial results.

A methodology must regularly make upfront investments to keep marginal cost low. Low marginal cost equates to financial agility and better financial results. Agile doesn't emphasize low marginal cost, so the Agile methodology doesn't feel financially agile.

#2: Tech-led Not Business-led

Agile is also misaligned with the ultimate goals of business. Since (to quote the manifesto) "working software is the primary measure of success," consider what takes a back seat: value propositions, profitability, and customer satisfaction. Agile's mission is for every two-week "sprint" to present new capabilities for the business to accept. Problems the business identifies are iterated (i.e., kicked down the road) to an 'eventual' future sprint.

Agile equates success with software—not revenue, profit, or customer satisfaction. When you execute Agile, software and technology seem to run the business when, instead, business should run the business.

#3: Documentation Debt and Traffic Jams

The third problem is that Agile emphasizes working software over comprehensive documentation. The Agile Manifesto shrugs at documentation that precedes building software. This reduces the clarity, transparency, alignment, and value of that information and the employees who document their collaboration, i.e., the business. Meetings and email fill the void, causing these communication channels to upstage documentation.

Agilists are intolerant of unnecessary documentation, but very tolerant of unnecessary meetings and ceremonies. Emphasizing meetings and emails creates meeting gridlock and email overload—a communication traffic jam. Months later, memories of meetings have faded, emails are buried, and the value of the work has depreciated. Hopping between back-to-back meetings might qualify as schedule agility; it's certainly communication traffic jams and schedule chaos. Meetings and email are wonderful servants, but horrible masters. Emphasizing them puts teams into documentation debt.

#4: *Culture of Neglect and Sprawl*

For many organizations, team culture continues to be a prominent and vital topic. It's clear that the Agile methodology shapes culture for software developers. What is less clear is Agile's influence on culture for other innovation team members.

Agile prescribes some specific events, but otherwise provides minimal guidance for the entire team. Therefore, it lacks the scope and rigor to do the whole job of shaping company culture. If anything, Agile encourages freedom and variability.

Verb sprawl™ is one form of variability where team members use exotic and expansive language to describe what are, in reality, a limited number of activities. But don't blame Agile; no one asked it to govern a working team or shape a working culture. The world only asked Agile to shape working software.

#5: *User Stories Don't Tell The Whole Story*

Agile's limitations with language are not limited to written verb sprawl and frequent meetings that depreciate. Unsurprisingly for a software-centric methodology, Agile does not leverage the power of storytelling. Agile mandates documentation called user 'stories'—however, they are typically data-centric, specifying content to retrieve or store in a database. Their focus on content neglects *context*—the bigger picture and the bigger customer story.

For innovation purposes, impactful storytelling captures the unacceptable 'current state' story, shapes the 'future state' story, and promises breath-taking moments that matter in your customer's story. Agile doesn't encourage innovators to deeply explore and expose the pain, pleasure, adventure, and delight in the long arc of the stakeholder story. Agile contains bigger bodies of work such as epics and features, but disproportionate attention goes to narrow user stories.

Finally, with the benefit of two decades of hindsight, it's fair to conclude that Agile fixed one symptom of Waterfall (big-bang projects) but not its problematic foundations. Both methodologies govern technology and fail to govern people. Both methodologies enable poor transparency by not enforcing current state documentation.

Exclusively reducing big-bang projects and pursuing small projects has encouraged and incentivized franticness. Healthy companies execute with a sense of urgency, but Agile lacks the structure and principles for entire innovation teams to collaborate with *sustainable* urgency.

Although cynical, Agile has another cultural difference from Waterfall. Project cancelation, failure, and changing your mind mid-project are not a big deal and, in fact, are considered a point of pride. Agilists call this "fail fast, fail forward," since the maximum waste is one sprint of work before the team gets a fresh start. In casino terms, Agile systemizes cutting your losses and normalizes disposable work.

The limitations of software-centric methodologies put companies at a disadvantage. This disadvantage is a kind of debt. Programmers experience 'tech debt' when shortcuts in code cause problems later. New employees experience 'documentation debt' when tenured employees can only pass on tribal knowledge. Entire companies have a disadvantage when they execute with a 'methodology debt.'

These all matter—cost profile, business-orientation, the right documentation, team culture, and effective storytelling. Agile limits organizations because Agile shrugs at these. The Elegance methodology aggressively pursues low marginal cost, prioritizes working teams over working software, strictly manages documentation, emulates a factory for a culture of discipline, and emulates the arts for storytelling. The Elegance methodology sets up your future stakeholders for success and is limited only by your imagination.

Nostalgia Undermines Innovation

Run to meet the future or it's going to run you down.

~ Anthony D'Angelo

Humans often reward tradition and default to staying within comfort zones. The practical form of this is called inertia. The sentimental form is called nostalgia. Whatever you call the forces to relive previous habits and culture, overvaluing it in your decision making undermines your innovation.

Humans value nostalgia. It's human nature to respect the accomplishments of the past. It's pleasurable to relive wonderful memories. But reverence for the past can go too far. There are vast differences between empathetically looking at the past (nostalgia) and empathetically looking toward the future (innovation).

A competitive, constantly improving world cannot live within the confines of nostalgia. Everything new, exciting, and profitable starts with empathy for the future—your future, your employees' future, and your customers' future. Nostalgia has its place, but a methodology meant to improve and build new customer experiences is not that place.

Resistance to change is a primary cliché in the dialogue about innovation. But when it's reframed as nostalgia, the risks of complacency become clear. Nostalgia can be a distraction and a drag on your environment. The status quo of every organization eventually becomes stale, obsolete, and unacceptable. Skeptics of new ideas are anchored to the past.

The downside of complacency even applies to your innovation methodology. Companies have made enormous investments in Agile in terms of processes and technology. The word agile is almost invincible, and the proliferation of the word in many business contexts reinforces the word's universal acceptance.

Ironically, although Agile practitioners caution their customers not to resist change, many Agile practitioners will resist change. For them, nothing will change until the status quo is more painful than transitioning away from Agile. However, the emergence of the term Hybrid suggests that a compelling alternate methodology could untether practitioners from their confidence, comfort zone, and investment in anything labeled 'agile.'

Change leaders who are sufficiently dissatisfied with Agile's limitations need to convert their resistance to change to an affinity for change—so their organization is forward-looking and forward-leaning. Organizations must turn a critical eye on their past and present to compete, improve, and survive.

To stay competitive, leaders must avoid nostalgia; it encourages investing in your past. Innovation demands that you invest in your future.

The Experience-Based Economy

People will forget what you said, people will forget what you did, but people will never forget how you made them feel.

~ Maya Angelou (1928-2014), American poet and civil rights activist

Over the past century, many businesses saw themselves as product, service, or solutions companies. But in the twenty-first century, many businesses now emphasize user and customer experience. All readers of this book live and work in an experience-based economy. The Elegance methodology is grounded in this experience-based economy.

In the past, the stereotypical company existed primarily to maximize shareholder value. But it is increasingly common for an organization to improve the welfare of other stakeholders as well: customers, employees, local communities, underprivileged populations, and humanity at large.

This shift away from shareholder welfare toward global stakeholder welfare is not solely about altruism, but the shift *is* about value and *values*. What consumers, customers, and employees value does not stay static. As values evolve, organizations that keep up and embody these values achieve a competitive advantage over organizations that shrug at the experience-based economy. While many books and methodologies have a strong orientation toward customer experience, methodology for the employee experience is playing catch-up.

The world is increasingly empathetic. Competitive and collaborative forces compel your organization to be the same. To create compelling customer experiences and value, your employees' contributions, your organization's purpose and profit goals require you to improve the stories of all your stakeholders.

This book aligns with this goal of positively impacting any and every stakeholder. The goal is to transcend project management's dry reputation and provide you with the tools to shape and govern customer, employee, and stakeholder experiences as ambitious as your organization dares. The Elegance methodology will guide you through building moments that matter.

A Culture of Innovation Elegance

A culture is strong when people work with each other, for each other.
A culture is weak when people work against each other,
for themselves.

~ Simon Sinek (b. 1973)British-American author and motivational speaker

The business world craves tools to improve company culture and live the values so often preached. The themes of the factory, asset portfolio, arts, and leadership shape culture and govern these desirable values. The methodology of Innovation Elegance offers culture disguised as a conveyer belt of team agreements, while at the same time encouraging culture disguised as your favorite performing art. Innovation Elegance promotes values such as self-confidence and humility—both disguised in documentation.

Poor methodology ruins projects, relationships, and jobs, while a good methodology inspires projects, relationships, and jobs. Therefore, methodology is the most effective tool for collaborating on valuable, enjoyable, and profitable work.

The best single word to encapsulate what the Elegance methodology confronts is 'messiness.' The acronym VUCA (Volatile, Uncertain, Complex, and Ambiguous) is the formal term for messiness. Innovation Elegance confronts VUCA by giving you the tools to perform elegant teamwork.

To combat *volatility*, the Elegance methodology's tools govern vigilance and responsiveness to market forces, foster elasticity with employee morale, and use stoplight ratings to monitor the health of team assets.

To address *uncertainty*, the tools help your team routinely collaborate on future priorities, perceived probabilities, questions, and risks.

To moderate *complexity*, the tools control variability, organize with simplicity, and hinder messiness.

To minimize *ambiguity*, the tools embrace scripts, information sharing, and a cadence for performance feedback.

Of course, the messy part of any organization is its people—specifically, how their ideas, goals, and incentives—compete. The cliché that every company wants competitive advantage has existed for decades. A new realization is that most companies under-engineer how to manage internal competition.

As the world moves faster, internal competition becomes more intense. At some point, hyper-competition causes gridlock and group paralysis. The era of profitable hyper-competition is ending because it makes innovation more difficult than it should be.

The frontier of innovation is now 'collaborative advantage.' Its empathy leads to inclusion, bigger markets, audiences, and profits. It gives clarity and confidence to do the proper work. It mutes competitive noise that is counterproductive to your organization's purpose. Innovation Elegance's people centricity unleashes your team's strengths, genius, and passion, while its discipline protects you from the patterns of poor methodology. Despite the phrase being a paradox, or perhaps because it is, a collaborative advantage like Innovation Elegance is the frontier of profitable innovation.

The Elegance methodology reimagines empathy. Empathy is not simply being nice or generous. Empathy-in-action is the ambition, tools, and ability to improve someone else's experience. This distinction is important because boldness in empathy is motivating and profitable. Boldness inspires speed, passion, and purpose. *Elegance* is discipline, empathy, and love disguised as a methodology.

The Elegance methodology aims to be simple enough for innovation beginners to put the basics into motion. For seasoned innovation professionals, the methodology provides building blocks absent from typical innovation teams. Skeptics of these building blocks—skeptics of documentation—often complain, "This is just more work!" But the Elegance methodology shows why certain forms of documentation are precisely the proper work.

As a change leader and innovation professional, it's time to embrace how your organization has outgrown a methodology that served you well for a generation. It's time to innovate how you innovate, with a methodology that sets you up for success.

How this Book Is Organized

For every minute spent organizing, an hour is earned.

~ Benjamin Franklin (1706-90), American inventor, scientist, political philosopher and diplomat

The presentation of the Elegance methodology occupies three books. This first volume discusses the factory and empathetic arts metaphors, explaining the why and how of the Elegance methodology. It's suited for innovation professionals at every point in their careers, but it is particularly useful for senior leaders who first want an overview without getting into the weeds.

The second volume, *Innovation Portfolio,* shares material for the asset portfolio metaphor. Approximately sixty assets get into the weeds—i.e., the what—of the methodology and the output of your agreement factory. It is best for detail-oriented professionals.

The third volume, *Elegant Leadership*, is a rigorous analysis of leadership—distinguishing, navigating, and neutralizing good, bad, and false leadership. It highlights how a few centralized leaders make organizations vulnerable, and how thus in the Elegance methodology every team member exercises well-placed leadership regardless of their title.

I hope the pages in this first volume inspire you to explore the others. Some material will be relevant to you now, some later, and some will educate you on your team members' work.

As your team builds and refines all the assets in the second volume, you'll see every culture trait of the factory improve your team's discipline. The asset portfolio is comprehensive; it includes project-independent and project-specific assets. It captures the lens of market players, the lens of employees at a team level, and the lens of employees individually. It captures process, people, and technology specifications. The number of assets and their manageable scope instill confidence and avoid paralysis for an innovation team. The asset portfolio is most valuable when a team treats it like a synchronized factory.

As your team builds and refines its asset portfolio, you'll also see every culture trait of the arts improve your team's empathy. The arts allow your team to be prescriptive (theater) and exploratory (improv). The arts distinguish leading versus following (partner dance). The arts accommodate hyper-collaborative (music ensemble) and hypercompetitive (martial arts) individuals. The arts show when leadership must set the table and when they must bang the table (parenting). The asset portfolio is most valuable when a team emulates the empathy and grace of these arts.

In this book, ruthlessness in the workplace doesn't encourage sweatshops, bullying, or toxic environments. The term only intends to convey *uncommon discipline*. Likewise, grace in the workplace doesn't encourage unconditional forgiveness and zero consequences. The word encourages *uncommon empathy*. Executing with ruthless grace combats VUCA and governs an elegant employee experience. Companies with a people-centric methodology transcend their software-centric peers. Companies that achieve Innovation Elegance transcend Agile with ruthlessness and grace.

If a culture of meetings and emails hinders you from doing so called real work, these metaphors and assets calibrate you toward that work—with extraordinary discipline and empathy. Previous methodologies emphasized software built by a team. The Elegance methodology emphasizes the discipline and empathy exercised by the team. Legacy methodologies demand working software. The Elegance methodology demands a working team. A culture of hyper-discipline (ruthlessness) and hyper-empathy (grace) boosts innovation literacy and the value of a team's real work and durable output.

You hold in your hand the language, habits, and culture of the next generation of tools for innovation success. Adopt this methodology in whatever sequence and speed is natural for you; adopt the tools that are easiest for you first. Start small but get started; the cost of delay is not trivial. Build momentum, habits, and culture to innovate how you innovate.

I am confident these metaphors and their cross-pollination will bring you and your team success. As you read, adopt, and execute the Innovation Elegance methodology, you will discover the same confidence.

RUTHLESS DISCIPLINE: CULTURE DISGUISED AS A FACTORY

Discipline is built by consistently performing small acts of courage.
~ Robin Sharma (b. 1969),
Indian-Canadian writer and business consultant

Your organization is a factory. The output of your factory might be pizzas, treated patients, mortgages, or baseball games. Even in e-commerce, your factory might be overnight stays, rideshares, or electronic payments. Every disciplined organization is managed like a factory.

Factory owners care about certain things that are easy to quantify, like economics, speed, waste, variability, and automation. They care about certain things that are subjective and harder to quantify, like quality, autonomy, and ease.

Your innovation work is also a kind of factory since different stakeholders—investors, customers, and employees—care about other dimensions of your factory:

- Customers care about your speed and their wasted time and effort.

- Employees care about wasted time, effort, and upcoming automation that impacts their job.

- Investors care about the money you spend on operations, firefighting, and the speed at which you innovate.

Executing like a factory brings clarity, simplicity, and transparency to your innovation work, operationalizing your values and culture.

This first half of the book discusses ten aspects of a factory and how your culture impacts your innovation factory. More importantly, it explains *why* you should manage innovation work like a factory.

We'll start with topics that have some urgency, such as economics, speed, quality, and waste, which often require short-term attention and decisions. Later topics like ease, autonomy, and elasticity require long-term attention and decisions. Each chapter paints a picture of a healthy factory and an unhealthy factory. The details contain the topics that arise in innovation decisions.

The recent increase in conversations about culture is well-intentioned, but most contain a lot of ambiguity; they only offer a vague notion that some culture traits affect your bottom line. They are detached from daily realities and need straightforward governance of behavior change. These chapters will connect your culture traits to factory traits and explain why minimizing culture traits that negatively affect your bottom line is important. They will give you the vocabulary you need to hold effective conversations about innovation decisions, whether that is with peers, subordinates, or superiors. Finally, these chapters contain precise tools that make adoption easier than retaining a messy status quo.

Across these chapters, the altitude of the content—the level of detail—varies according to what applies to project professionals. However, it matches what is necessary to combat VUCA and address cultural problems common in teamwork.

Acute attention to detail can be unappealing to some innovation professionals. If a topic feels wildly detailed to you and 'in the weeds,' it means you are vulnerable to a culture trait that the section aims to solve. Share the material with a colleague who sees its value.

As this section on the factory metaphor digs into detail, recall that the title includes the term 'ruthless.' To conquer the powerful forces of VUCA, you need powerful tools. You need ruthless tools. Ruthless discipline brings health, value, and success. Instead of messiness, you will achieve elegance.

Economics

A business without a path to profit isn't a business, it's a hobby.

~ Jason Fried (b. 1974), American author and entrepreneur

The factory metaphor isn't for the sake of being cute or creative. The metaphor aims to optimize the economics of your innovation team by describing what good and bad factories look like.

Economically, your innovation succeeds when you:

- Set a realistic, modest scope, then optimize for that scope.

- Apply the proper rigor (and short leash) to minimize surprises.

- Prioritize by starting the right projects, in the right order, at the right time.

- Pause and pivot when new information appears.

Executives, investors, and Boards of Directors understand failed projects, failed investments, and high project failure rates. They hear about failed change and failing *to* change. The cost of a single failed project is high, but it pales in comparison to failure across projects due to poor methodology. These costs are exponentially higher.

Your innovation organization is heads-down on numerous projects. Innovation leaders must be heads-up on project economics to pause the less valuable work and advance the most valuable work.

Set and Optimize Scope

When eating an elephant take one bite at a time.

~ Creighton Abrams (1914-1974), American military leader

Every innovation impacts a specific population of stakeholders. A critical early step is correctly identifying that population and then fully serving them. When costs and benefits (real or perceived) are not uniform across stakeholders, tradeoffs arise, and decisions get difficult.

Serving too small a population is called 'optimizing locally.' Optimizing locally risks outsiders feeling that they pay more for what they receive in benefits. They may feel that decisions are unfair, or that their input subsidizes insiders' benefits.

In comparison, 'boiling the ocean' refers to work that serves too great a scope or population than what you can realistically manage. It overpromises and damages the experience for all stakeholders. It's good to always keep the bigger picture in mind, but managing too big of a picture and boiling the ocean is bad.

The ideal scenario is the selection and service of a respectable population of stakeholders. This is 'optimizing globally.' Optimizing globally allocates costs and benefits among stakeholders as fairly as possible. Allocation is subjective and depends on the values of each stakeholder. Therefore, optimizing globally requires countless tough decisions.

The following table shows these three scenarios (optimizing locally, boiling the ocean, and optimizing globally) for three different stakeholder groups.

Stakeholder list	Optimize locally	Boil the ocean	Optimize globally
Family with four children choosing a restaurant	Choosing a restaurant that only one of the four children likes	Ordering dinner from four different restaurants creates a headache for everyone	Choosing a restaurant that has enough options to satisfy each child
Company with lines of business in banking, insurance, healthcare, government	Build a unified process that accommodates the slowest LOB	Custom innovation for each LOB is slow and disappoints all LOB stakeholders	Build dual processes to accommodate high and low-friction LOB
Professional sports league of thirty teams	Each team's profits mirror the size of their market. Small markets have small profit. Large markets do not share their large profits, making the league barely attractive for teams in small markets.	Supporting more teams than what is optimal profitability for the league.	Supporting the number of teams that optimizes profits across the league and revenue sharing among all teams

The Optimize Locally and Boil the Ocean columns show short-sighted decisions. The final column shows tough decisions—signs you are optimizing globally.

The controversial lens on setting and serving scope is where mutual accountability exists. When economics favor collective responsibility, set a large scope. When economics favor independence and no accountability, set a small scope. Your organization's boundaries, interdependencies, and empathy for other organizations shape what matters to you for both short and long-term decisions.

For your organization's innovation, avoid optimizing locally or boiling the ocean. These are prone to poor value and poor profitability. To optimize financial results, optimize globally. Set your scope to the correct population and serve them well.

Manage With a Short Leash

To listen closely and reply well is the highest perfection
we are able to attain in the art of conversation.
~ François de La Rochefoucauld (1613-1680), French author

Optimizing globally does not mean managing from a distance. Projects with a long leash get into trouble. To minimize negative surprises, manage with a short leash.

The first way to manage with a short leash is to limit the duration of individual projects. Recall that the Waterfall methodology fell out of favor primarily because go-live events were far apart (nine to eighteen months). But go-live events don't need to be two weeks apart, either. Projects with a modest duration have a short leash and keep risk low. Instead of the extremes of two weeks or two years, consider a project length of nine to eighteen weeks.

The second way to manage with a short leash is to set interim milestones and formalize the success of every milestone before proceeding. This keeps failures fast and small, minimizing rework.

The third way is to stay loyal to the value proposition of the investment. Because teams commonly wrestle with teamwork and project mechanics, they 'take their eye off the ball' and lose sight of stakeholder expectations. Healthy innovation teams minimize ambiguity and maximize traceability through every project asset, keeping them close to the original value proposition.

There is no upside to a project with a long leash. A long leash is taking on a lengthy project, treating milestones casually, and losing sight of the value proposition. Sound economics demand that you manage with a short leash.

Prioritize and Pace

Most of us spend too much time on what is urgent and not enough time on what is important.

~ Stephen Covey (1932-2012), American author, businessman, and motivational speaker

If you have sized your projects modestly, you have the right projects. Next, sequence and pace them according to priority, size, and team intensity. Every project's value proposition narrowly assumes optimal timing, but optimizing globally requires pacing across all approved projects.

One prioritization approach skips calculations and crudely weighs impact and effort. High-impact and low-effort projects receive higher priority, of course. For some organizations, this can work.

When financial calculations are needed, prioritize projects with the highest expected stakeholder value and company profitability. Expected profitability depends on three components:

- Impact on ongoing revenue (price times quantity).

- Impact on ongoing cost (variable or marginal cost).

- Cost of the investment itself (fixed or upfront cost).

An innovation is a good idea if it (on its own or as part of a bundle) is among the options with the highest expected profitability. As you decide the number of projects to start, have a firm stance on what project is the next most valuable (NMV). When not bundling projects, knowing the NMV project helps to queue your prospective projects.

Companies typically do not plan or execute projects one at a time. They also avoid lumpiness, i.e., dramatic swings in the number of projects simultaneously in-flight.

Thoughtful innovation leaders know their optimal intensity. They start and execute projects at a sustainable pace. This translates to a target number of simultaneous projects and the avoidance of too many projects starting or ending in a short period.

On the topic of timing, avoid starting a project too soon. A company is vulnerable to this if the business case lacks rigor, accountability, or is too optimistic. Premature projects are often pet projects for individuals with too much power and influence. These are impulsive ideas originating from a shiny object—something catchy from a distance, but disappointing in substance upon a closer look. Sadly, it's common to start a project when a pragmatic value proposition would have concluded not to perform the project at all.

Of course, starting a project too late is also something to avoid. Procrastination makes your stakeholders feel neglected. They might proceed without you (a form of optimizing locally) or partner with a competitor. A value proposition might be time-sensitive, i.e., the value to your stakeholders and expected profits could disappear.

To avoid innovating too early or too late, start projects at their tipping point. For a single project, the tipping point is when the status quo is more painful (or less profitable) than the transition. For a group of projects, the tipping point includes the perceived 'least profitable project' that *is* still profitable. Start prioritized projects at the time that minimizes the cost of delay, the loss of profit, and the obstruction of more valuable projects already in-flight.

Another consideration for the tipping point is the number of items involved in a process: i.e., a customer or employee experience. When a street with increasing traffic reaches a tipping point of the number of cars per hour, the economics justify a traffic light. When a growing company reaches a tipping point in the number of employees, the economics justify a corporate cafeteria to serve lunch.

For a company that reaches a tipping point of innovation intensity, the economics justify an innovation methodology. These economies of scale matter in investment decisions because they reduce marginal cost. The ideal prioritization is that projects, individually or as a bundle, start in a thoughtfully timed, paced, and transparent manner.

Pause and Pivot

Pausing is not the end of the disruption process, but the beginning of the next leg of your journey.

~ Jay Samit (b. 1961), American author and digital media innovator

In this dynamic economy, an innovation team may learn something new mid-project. When this happens, a skilled innovation team is ready to pause and pivot.

One scenario is that the team learns something new that originates within the company. Personnel changes, acquisitions, or divestitures can cause sudden changes in relative value propositions and may become reasons to stop a project.

Alternatively, the team might learn something new that originates outside the company. Regulation, competition, or a crisis can disrupt in-flight innovations. And of course, your customers might change their minds about their interest in an innovation you have in motion. Positive or negative news might warrant a difficult, but disciplined, decision to pause a project and pivot.

Don't make irreversible decisions or tolerate a project to languish. Procrastination makes pivoting more difficult. Inattentiveness can be devastating. It's possible and essential to stay attentive, decisive, and able to reverse.

New information might require quick action. In a crisis where physical safety is at risk, skip the math. If it is not a crisis, the business case rigor is worthwhile even though value propositions shift. New information isn't always bad. New information might contain new markets, new customers, and new career opportunities.

A project pause might cause feelings of wasted work or negative morale. Pausing is reversible, and disciplined documentation eases resuming the project later. Pausing and resuming are inexpensive pivots.

Good innovation leaders are attentive to new information. They respond pragmatically and with humility. Inattentiveness and a lack of humility cause negative surprises and, in the worst-case scenario—a project delay—an expensive pivot.

Moving The Goalposts

Humanity's special place in the cosmos is one of
abandoned claims and moving goalposts.
~ Frans de Waal (b. 1948), Dutch primatologist and ethologist

One sad, painful cliché of innovation is a project that encounters trouble, resets expectations a few times, and continues working toward a diminished value proposition. This resetting of expectations created the sports analogy: moving the goalposts.

Keeping the size of your projects modest limits the downside of any negative surprises. This on its own limits the distance of any goalpost moving. But there are occasions where moving the goalposts dramatically impacts the value proposition of an investment. In an extreme case, moved goalposts erase the expected profitability altogether. But more likely is that one of the following six scenarios moves your goalposts.

Timing	Size of Surprise as a Percentage of Original Budget	Damage Control Level
Early–10% into the project	10%	Low
Early–10% into the project	50%	Low
Middle–50% into the project	10%	Low
Middle–50% into the project	50%	Medium
Late–90% into the project	10%	Medium
Late–90% into the project	50%	High

The table's first two rows represent incidents near the start of the project, i.e., 10% of the schedule burned. Regardless of the size of a schedule reset (represented by the middle column), changes are not a big deal, and the need for damage control is low (represented by the right column). These are examples of *failing fast*.

The middle two rows describe an incident halfway through a project. The fifth row *fails small*, but a negative surprise this late is a case of *fail slow* instead of *fail fast*. The sixth row is a case of *fail big* instead of *fail small*, and thus the need for damage control is greater.

The table's last two rows are near the project's perceived end: perceived, since doubling a project timeline undermines that claim. The project might have *burned* 90% of the baselined schedule, but doubling the schedule means the team believes they have *earned* 50% of the project's value-added work.

This change is a big deal, and the need for damage control is high. This is expectation management at its worst.

Expectation management, perception management, and damage control all matter. Organizations certainly prefer to learn about trouble sooner rather than later. But something else to consider is whether the organization would make a different *decision* if they knew then what they know now. When investment prioritization is sensitive to schedule, investment decisions might be different, and the risk for damage control is high. If prioritization is not sensitive to a schedule delay, then the risk for damage control should be low.

As the twenty-first century marches on, innovation complexity increases, transparency increases, and the notion of 'on time and on budget' is less meaningful. Organizations rarely report, "Here at 5% completion, our effort and duration estimates remain accurate, precise, and stable!" The notion of a static schedule, scope, budget, and effort is counterproductive and unrealistic. Instead, a healthy and honest innovation approach is "Prioritize and go!"

The Elegance methodology details the assets that reduce negative surprises, the magnitude and likelihood of moving goalposts, and the need for damage control. If a project moves the goalposts, your asset portfolio is critical to fail fast, fail small, and fail forward.

Forgiving End to a Successful Project

True freedom is impossible without a mind made free by discipline.
~ Mortimer Adler (1902-2001), American philosopher, educator, and author

The economics of a project are forgiving either at the beginning or at the end of the work—never both. Methodology debt and leadership voids result in a forgiving start and an unforgiving end to any project. A forgiving start often leads to surprises partway through a project. When a project subsequently refuses to move the goalposts, the project team has to firefight to meet the original schedule. The firefighting leads to finger-pointing (instead of healthy accountability), personality conflict (instead of task conflict), and demonization (instead of disagreements). The long hours to meet this go-live date are the unforgiving end to a project.

However, a robust methodology translates to an unforgiving start and a forgiving end.

An example is a new boss asking a project manager about a recently completed project.

Boss: What do you mean? You don't remember the exact go-live date? You don't remember when the project ended?

PM: Instead of a single date, the go-live contained a handful of activities staggered over a few days in the second week of October. The team built a very robust schedule that exposed every change along the way. It had all the right ingredients to avoid big problems. It could handle minor delays or changes. Accountability and ownership were strong; the team was self-sufficient at that point. Unless I stepped away, I would have been unnecessarily micromanaging. That was around October 1st; the team did not need me for the last two weeks. We completed a 'lessons learned' exercise before the end of the month, which documented over a hundred comments across the team, and we went into maintenance mode. The training was well-received and the go-live wasn't a crisis. I think it's great it was anticlimactic.

This conversation reflected a project that finished well. The upfront planning was rigorous, transparent, and continually vulnerable, and everyone had access to detailed documentation and team meetings. This meant the negative surprises were few, and the team had the resources to respond to them along the way. Instead of moving the goalposts, the team waltzed across the goal line.

To achieve the original value proposition, you must be unforgiving at the start of the project. This discipline pertains to scope, priorities, pacing, and the ability to pause and pivot.

Like many forms of discipline, this upfront rigor is often unpopular. Just remind yourself and your team that ruthless discipline enables a graceful, forgiving end to projects. Like all methodologies, its ingredients are not just for the sake of methodology. Innovation Elegance is about value, economics, and money.

Speed/Tempo

The pace of change in our world will never be slower than it is right now.

~ Beth Comstock (b. 1960), Chief Marketing and
Commercial Officer, General Electric; author of *Imagine It Forward:
Courage, Creativity, and the Power of Change*

E very factory has an optimal speed. A factory does not need to run constantly at that speed, but it's important to know what that speed is and to execute close to it. In addition, every *part* of a factory has an optimal speed—not for its own sake, but in relation to other components. The synchronization of interdependent parts regulates the speed of the bigger picture.

Drags on Speed

*Good teams beat you with speed. Great teams beat you
with spacing and timing.*

~ Jason Kidd (b. 1973), American professional basketball player

Although factories can go too fast, the more common problem is that culture traits make factories too slow. Culture traits are forms of friction. You want to minimize unhealthy forms of friction since they slow your factory. Slow work disappoints stakeholders and delays revenue.

Bureaucracy

Bureaucracy is one form of unhealthy friction, and it hurts speed. Bureaucracy is the formal term for too many cooks in the kitchen. Bureaucracy is transparency and inclusivity gone too far.

Overcorrecting for bureaucracy is a different issue—a silo. Silos also cause less-than-ideal speed due to neglect, exclusion, or impatience with contributions from valuable stakeholders. Silos optimize locally; to optimize globally, silos should slow down and involve more stakeholders.

Ambiguity

Ambiguity hurts speed. Some ambiguity is avoidable; some is not. It's a valuable skill to be *able* to work with ambiguity, but another valuable skill is reducing and eliminating ambiguity for others. Workers who lack clarity look for clues on how to do their best while uncertain about their work's purpose, quality, and accuracy.

One form of ambiguity is 'documentation debt,' where an organization neglects documentation. This debt increases tribal knowledge, which is information that only resides in the heads of a small number of employees and leaves the company when those people do. Tribal knowledge makes people indispensable, introduces bottlenecks in teamwork, and decreases everyone's mobility.

Another contributor to ambiguity is a trendy approach to 'manage by outcomes' (MBO). Of course, results matter, but emphasizing lagging indicators and shrugging at leading indicators is neglectful. MBO grants so much freedom that it effectively says that *how* work gets done doesn't matter. MBO embraces ambiguity and contributes to VUCA. It refutes the belief that every big success requires 1001 small successes. Small successes are leading indicators and confidence builders. The transparency of small successes minimizes ambiguity.

Low ambiguity about your freedom helps speed. It is valuable to know when work is mandatory or optional, and whether it is on-script (prescriptive) versus off-script (improvisational). Low ambiguity is forthright, straightforward, and improves speed.

Complexity

Complicated and complex things hurt speed. These terms are often confused, but both hurt your speed. A complicated situation has numerous parts, perhaps organized poorly. But if you solve its puzzle, you can formulate a recipe and achieve some control and certainty.

A complex situation has variables, unknowns, and interdependencies that obstruct a clear resolution. In complex situations, a recipe is less helpful and can even be counterproductive. Complexity is a silent killer of speed, whereas simplicity aids speed.

It's easy to confuse the terms 'simple' and 'simplistic.' Being simple means being thoughtful, mindful, straightforward, and easy; being simplistic is being thoughtless, careless, and neglectful.

Interruptions

Interruptions hurt speed. Time dealing with interruptions is called 'switching cost.' It requires time away from the tasks at hand. The ability to be interrupted and spontaneous is a strength, but frequently being interrupted is a disadvantage.

Fatigue

Fatigue slows work down. Speed suffers when a key contributor is overloaded for a significant period. If they reach the point of burnout, their speed is zero, and they are a bottleneck for the entire factory. To keep your organization sustainable, prevent anyone from taking on too much. It will also prevent your colleagues from slipping into a victim or martyr mentality.

Flooding

Flooding reduces speed. When many objects are in motion simultaneously, gridlock forms and everything dramatically slows down. Synchronizing takes a back seat to teams navigating the flood until the gridlock dissipates.

Information overload is a form of flooding. The overused 'reply all' email function is one example of this. Another is the trendy phrase, "I need a 360-degree view of the customer." This translates as: "I can't decide unless I have 100% of the information," when of course you *can* and often *do*. Navel gazing and analysis paralysis slow progress. Know enough to move on, align with your customer on the most valuable work, then get going by focusing and executing.

Perfectionism

Perfectionism hurts speed. Instead, Oklahoma-based pastor and author Craig Groeschel encourages leaders to embrace and execute GETMO: "good enough to move on." Most activities reach a point where an incremental hour of effort is better spent on something else. All work has a point of diminishing returns given that somewhere along the path to perfection, other work becomes more valuable.

Ego

Egos slow progress. For example, consider a person with a hero mentality who is focused on themselves rather than customer-centric work. They extract disproportionate attention, credit, and decision rights. They monopolize opinions and create bottlenecks. They position themselves as single points of failure, which jeopardizes team continuity, sustainability, and value. This person's colleagues inevitably allocate more time and attention to this person's ego at the expense of real, customer-centric work, hijacking the role of the customer without the revenue.

Another form of ego that impedes speed is a personality type called a contrarian. A contrarian's typical behavior is to give contrary views much more frequently than is expected or helpful. A contrarian does this to get attention, undermine a specific individual, or undermine the entire effort. A contrarian injects unpredictability, chaos, and politics into the work. The discussion of leadership in the third Innovation Elegance book specifically includes ways individuals and companies can address these ego-based problems.

Getting Up to Speed

The synchronization skill of the boss is the speed of the team.
~ Paraphrased quote, Lee Iococca (1924-2019),
American automobile executive

Culture traits that boost speed include power, influence, focus, and trust.

Power takes forms such as rank, rights, responsibility, empowerment, and ownership. These forms of power reduce friction, avoid stagnation, and enable work to proceed.

Influence takes forms such as persuasion and negotiation. Influential team members are patient with their colleagues, share rationales, and align on decisions so that teamwork can move on.

Focus equates to knowing how to *not* spend your time. It requires saying no to numerous alternatives. Inspiration and obsession are the extreme forms of focus. Achieving focus depends on later discussed topics such as waste, variability, and autonomy.

Trust

High trust accelerates progress. Poor trust slows work down. Trust takes different forms (examples below) so attribute trust's impact on speed to one or more of these forms. To improve speed, explore the sources of mistrust and identify how to manage differently.

If there is distrust in someone's _____	Someone else spends more time _____
Competence	inspecting their work
Integrity	monitoring their activity and ethics
Dependability	verifying their commitments and progress
Benevolence	worried about themselves more than doing their job

One way to envision speed uses the analogy of a mountain climber. A novice mountain climber, valuing speed, charges up the hill without studying the climb. They do not see the time and safety benefits of studying a climb (a high upfront cost) before proceeding.

An expert climber accepts the upfront time, energy, and cost; they invest in studying the path before starting the climb. An expert climber does the climb in their mind first so the actual climb avoids high cost in the form of delays. They 'go slow to go fast,' or, in the words of the military proverb, "Slow is steady. Steady is smooth. And smooth is fast."

Label the culture traits that cause your factory speed to be unsustainably fast or disappointingly slow. These traits cause unhealthy levels of friction. Like an expert mountain climber, be thoughtful how you can increase speed. Minimize the culture traits that hurt individuals' speed, then synchronize your team's many moving parts.

Quality

No one changes the world who isn't obsessed.

~ Billie Jean King (b. 1943), American tennis legend and gender rights pioneer

Most people would agree that the quality of factory output matters. But quality is subjective, and people can disagree about what constitutes high quality.

What constitutes quality in innovation (versus quality in operations) is always open to interpretation. Disciplined innovation professionals shape their culture to produce experiences that customers and stakeholders consider high quality.

Operationally, a high-quality item has unique features that appeal to the five senses: sight, sound, smell, taste, and touch. A high-quality item shows exceptional attentiveness to customer needs and wants. Special attention to detail results from deep listening, caring, and serving.

These traits of high quality translate to innovation. An innovation team's output must attend to the needs and wants of customers and stakeholders alike, and reflect focused listening, caring, and serving.

The focused path to high innovation quality starts with a firm command of what your organization does to uniquely benefit its paying customers and stakeholders. This is your organization's 'why'—its *purpose*. It's difficult to lead without a sense of purpose, and it is difficult to follow someone who lacks a sense of purpose.

Everything your organization does should trace back to this sense of purpose. Traceability formalizes the connections between your detailed work and your lofty strategic goals. It verifies that all details contribute to the organization's purpose—nothing is extraneous, and nothing dilutes the organization's distinct mission. Traceability aims to eliminate orphans among strategic goals and in detailed work.

The work cultures of humans are imperfect, so achieving complete traceability and ruthless focus is difficult. Several common culture traits can impact innovation quality because they undermine how traceable your company is to its purpose. But when your culture is obsessed with your purpose, there are no limits to your quality.

Culture Traits That Impact Quality

If you want to go fast, go alone. If you want to go far, go together.

~ African proverb

Speed and quality are highly related. Some culture traits improve both, but as you reach an optimal level of either, pushing even more to improve one eventually hurts the other.

As discussed earlier, ambiguity and bureaucracy are culture traits that hurt speed, and GETMO is a culture trait that helps speed. In that same vein, perfectionism, fear, silos, boiling the ocean, favoritism, and loyalty are culture traits that hurt quality, while fairness is a culture trait that conveys quality.

Perfectionism
Quality gone too far is perfectionism. A typical worker strives for high quality, but at some point, their incremental work is less valuable than the time the next stakeholder waits for the handoff. Perfectionism hurts quality.

Fear

Fear hurts quality. It is human nature to fear the exposure of our mediocrity and dysfunction, and to fear the disruption to the status quo. But optimizing quality requires overcoming these fears.

Fear suppresses ideas, feedback, and resourcefulness. Fear undermines talking and listening, creating blind spots. Vulnerability on your own terms is healthy, but vulnerability on someone else's terms is fear, and that hurts quality.

Franticness, which is reckless speed based on fear, hurts quality. Many innovation professionals are addicted to franticness because they believe a veneer of anxiety proves their value. Trade your anxious culture for poise. Trade the fear of change for a fear of standing still.

Silos

Operating in a silo hurts quality. Silos err by having too few contributors, inspections, or recipients of the work. Silos optimize locally at the expense of the bigger picture. They generate surprises and rework because of too *few* cooks in the kitchen and too *little* organizational friction. Silos gave birth to the clichés 'penny-wise and pound foolish' and 'win the battle but lose the war.'

Boiling the Ocean

Employees and their teams have limits and boundaries; disregarding these limits and boundaries hurts quality. Avoid boiling the ocean with your ambitions. Be realistic about your capacity; be decisive; and be at peace with completing work that stays high quality.

Favoritism

Society sees loyalty and favoritism as positive things—forms of subjectivity, reliability, and good judgment. But in an innovation team, they hurt quality. Both reflect irrational deference to letting past investments and activity influence your current decisions. Both risk making you a victim to the 'sunk cost fallacy' of overvaluing the past.

Overvaluing past loyalties and playing favorites within a team marginalizes others, diminishing engagement, morale, and the contribution of most of the team. Favoritism concentrates benefits on a lucky few employees, customers, or stakeholders.

Fairness

A culture trait that improves quality is fairness. Fairness in your culture is a form of optimizing globally. Fairness breeds the best ideas and performance. Fairness fosters high-quality teamwork.

Fairness also applies to competition between people. If the losers feel they lost a fair competition, they will reflect on how to compete better. If the losers feel they lost an unfair competition, they won't compete again. Loyalty and favoritism undermine process fairness. Extreme loyalty resists change and undermines innovation.

Fairness isn't free. It has cost. Fairness has an elegant, scalable cost profile of a high upfront cost and low marginal cost. Upfront costs for fair competition include rules, governance, standards, visibility, and transparency. Low marginal costs include judges' work, awards (for winners), celebration, and reflection (performed by everyone).

Opponents of fairness dislike its upfront costs. An unfair culture prefers competition with ambiguous rules, double standards or no standards, and low transparency. Unfair competition creates high marginal costs such as complaints, audits, investigations, damaged relationships, and lawsuits. With high marginal cost, unfair competition hurts economics and attracts poor talent and low effort. Unfairness hurts quality.

Career Security Over Job Security

Work harder on yourself than you do on your job.

~ Jim Rohn (1930-2009), American entrepreneur, author,
and motivational speaker

Consciously or not, every worker makes choices related to their job security and career security. To optimize quality, a healthy culture promotes career security over job security. Choosing job security is short-sighted, while choosing career security optimizes your big picture because it keeps your future customers in mind. Otherwise, if you don't listen to the bigger market, you create blind spots and unhealthy vulnerability.

It's natural to fear losing your job. But that fear negatively impacts teamwork. Fearful employees limit how helpful they are to others. They are prone to marginalize and undermine other team members. They want to be an indispensable single point of failure, and they cannot easily move out of their role.

A job-centric person is stuck. They resist change and innovation and make themselves dependent on people currently in power. Instead of being in control, job security becomes a trap. Job-centric employees hold others captive and hurt innovation quality.

The twenty-first century reality is that jobs come and go. In fact, entire job categories come and go. Neither is safe in the long term. A career-centric person embraces fluidity in organizations and knows change can come at any time. Although it can feel less secure in the short term, career security breeds resilience and sustainability for long-term employment.

Organizations have a financial incentive to have career-centric employees. These employees fearlessly work themselves out of a job and delegate to cheaper junior employees. With a low level of fear, they are not dependent upon a static power structure. Instead, they invest in their skills, network, and future colleagues.

These kinds of employees are curious about the future and welcome its arrival. They reinvent themselves on their own terms and are ready when a promotion opportunity appears. Career-centric employees improve innovation quality.

Business and innovation are all about matchmaking. A person fills a job when all parties feel they're a good fit—a good match. A culture of career security places people where they are most valuable. Thus, a commitment to career security requires you to play matchmaker. This commitment inspires the courage, discipline, and humility to separate a person and a job when both have a better fit elsewhere.

Task Conflict Over Personality Conflict

Creativity comes from a conflict of ideas.

~ Donatella Versace (b. 1955), Italian fashion designer
and businesswoman

High quality in innovation requires diverse skills and diverse ideas. Diversity is susceptible to disagreement, and disagreement can erode into personality conflict. Although personality conflict undermines quality, task conflict enhances quality. Great cultures have tools to pursue diversity, minimize personality conflict, and welcome task conflict.

Personality conflict happens when people distrust and demonize others. Good leaders squash personality conflict. As cartoonist Scott Adams of *Dilbert* fame once said, "Everyone is someone else's weirdo." Remind colleagues that unfamiliar language, habits, and culture appear weird to everyone at first. Celebrate the new and get to work because diversity in collaboration always results in superior outcomes. Heterogenous teams outperform homogeneous teams.

Note that the absence of conflict is not harmony; it is apathy. Without conflict, teams lose their effectiveness. Task conflict is healthy and so inevitable that establishing a decision process to govern debates makes sense. Process fairness is enormously important to most people; they are willing to accept outcomes they dislike if they believe the process that led to the results was fair.

Good leaders evangelize healthy task conflict. Conflict doesn't, and shouldn't, include hostility. Disagreeing is okay. Demonizing is not.

Sometimes overcoming a personality conflict is as simple as having argumentative individuals sit on the same side of a meeting table. A conference room setup and what people must look at during a meeting can influence body language and any tendency to be confrontational.

A more inconspicuous form of conflict is silent dissent. Leaders must probe for silent dissent because something might be important to hear. Extended silent dissent often involves fear, bullying, and oppression. It can motivate sabotage.

Whatever form it takes, when team members engage in personality conflict, their attention and criticism are in the wrong place. A healthy team directs criticism toward documentation content instead of toward people.

Innovation Elegance is obsessed with documentation because it encourages task conflict and undermines personality conflict. Without documentation to build and review together, team members must look at each other. The absence of documentation harbors personality conflict, enabling counterproductive confrontation or acquiescence among people. Documentation is a clear vehicle to capture team members' attention, contribution, and frustration. A commitment to welcome task conflict and prohibit personality conflict improves the quality of your innovation work.

One tactic to encourage healthy conflict and avoid polarization is to increase the number of options. Conflict is more vulnerable to turning personal when debating between only two options. But task conflict needs to be moderate and moderated, and one way to do that is with more options.

Stalemate or an impasse are signs that a team misaligned somewhere in the past, but only discovered that now. The quality problem resides in an upstream asset, and the team must return to it. Re-aligning on the upstream asset enables resolving the impasse within the downstream asset.

Unusually low conflict is a sign of acquiescence, apathy, or incompetence. Moderate conflict is a sign of competence, empathy for other perspectives, and the courage to defend yourself.

Competition of ideas is valuable and inevitable. Ideally, your culture anticipates this competition, and winning ideas are traceable to your organization's purpose. Competition of *people* within your team is not traceable to your organization's purpose. Instead of a boxing ring of people, host a boxing ring of ideas. Anyone unable to behave in the boxing ring (i.e., personality conflict) needs to leave the boxing ring. Those who stay in the ring embrace moderate task conflict.

Durability and Five Verbs®

The great use of life is to spend it for something that will outlast it.
~ William James (1842-1910), American philosopher, historian, and psychologist

The last aspect of quality to consider is durability. Outside your work culture, you might think of durability as a twenty-year-old car, a forty-year-old professional athlete, or a 300-year-old piece of music still relevant today.

Inside your innovation culture, teamwork results also need to have durability. The raw output of meetings and email lacks durability, which is why the Elegance methodology emphasizes documentation. A person averse to documentation undermines durability and lowers the value of the work they influence. For optimal durability, build the right documentation, in the right sequence, with the right contributions from the right stakeholders. This ensures the team asks and answers questions at the right time. Documentation built this way retains its value across projects. It creates an asset portfolio for your innovation team to refer back to as needed.

To ensure the quality of teamwork assets, adopt a reliable process and make the process a habit:

1. First, a stakeholder drafts the asset.

2. Second, some combination of stakeholders reviews it.

3. Someone (likely the drafter) revises it.

4. Some combination of stakeholders approves the asset.

5. And finally, someone distributes the asset to non-contributing stakeholders.

The Elegance methodology calls this framework, Five Verbs. The five verbs are: draft, review, revise, approve, and distribute.[4] Five Verbs shapes the decision process to govern debates.

This structure—especially the word choice of Five Verbs—might feel administrative and petty at first. But this is a case where, in the words of W. Clement Stone (a Chicago-based businessperson and philanthropist), "small hinges swing big doors." Five Verbs is a habit. A habit is shared behavior. Shared behavior shapes culture.

The discipline and rigor in Five Verbs are ruthless. Assigning team members to the Five Verbs framework is simple and straightforward, enforcing an obsession with quality. This ruthlessness is how an agreement factory transcends Agile and combats VUCA.

The durability of the Elegance methodology's agreement factory is two-fold. First, documentation has higher upfront costs than meetings and emails, but ongoing (marginal) costs to revise and educate on the content are low. At a higher level, overcoming inertia to adopt this agreement factory has upfront costs, but when a team stops reinventing the wheel, marginal costs to govern the work plummet.

Embracing this 'small hinge' simplifies mechanics and significantly liberates a team's brainpower from administrative planning (and replanning) to work on the actual, durable team output. This is 'swinging a big door.' Instead of rehashing verbs for disposable work, continually execute the Five Verbs for durable work.

---

4   Five Verbs resembles and improves upon a common project management framework: the RACI matrix (Responsible, Accountable, Consulted, and Informed). Assigning; who is responsible → Draft; who is consulted → Review and Revise; who is accountable → Approve, and who is informed → Distribute. A RACI matrix has noble intent but harbors ambiguity and lacks teeth. Managing via the Five Verbs framework has low ambiguity, with the teeth to build a durable asset portfolio.

In addition to enabling durable work, Five Verbs minimizes the marginal cost of three non-durable culture traits: meeting gridlock, email overload, and replanning. Every day that you relax or apologize for being obsessed about quality keeps your marginal costs high. Procrastination reduces your durability, quality, value, and profit.

Conventional thinking says quality is subjective. But these culture traits are universally attractive. Pursuing them unquestionably optimizes the quality of teamwork.

Waste

If you don't know where you want to go, it doesn't matter which path you take.
~ Lewis Carroll, Alice in Wonderland

I n a factory, a modest amount of waste is inevitable, and certainly not a crime. What is a problem is a culture that continually generates a lot of waste. Change leaders owe it to their teams, customers, and stakeholders to shape a culture that keeps waste levels low.

For a conversation about waste in your innovation culture, the best starting point covers the basics of waste. The topic's definitive authority is Toyota Motor Company. The auto manufacturer defines eight forms of waste. The table below gives a snapshot of team waste. It maps each type of waste to the communication channels that all innovation professionals know well, i.e., meetings and email.

Type of Waste	In Meetings	In Email
Overproduction	Over-schedule, over-include, over-attend, long-winded, duplicates	Numerous 'reply alls' Adding to a long string of messages
Waiting	Vulnerable to schedule availability and last-minute cancelations	Backlogged inbox, overlooked or mis-prioritized email
Inessential handling	Unnecessary attendees Unnecessary topics	Unnecessary sending, receiving, reading
Non-value-added processing	Rehashing conversations Forgotten conversations	Redundant content Off-topic content
Inventory in excess of immediate needs	Topics in the wrong order	Premature information
Inessential motion	Extraneous commentary	Extraneous commentary
Correction caused by defects	Everyone who needs to know will not remember in a few months	Agreements reside across multiple emails
Unused talent	Inauthenticity Someone being quieter than they otherwise would	Inauthenticity Someone being quieter than they otherwise would

Day-to-day, the amount of waste in meetings and email seems mild. But week-to-week, cultures with poor meeting and email habits generate a lot of waste.

Some forms of waste are less day-to-day but are just as common. An entire project can be prioritized poorly or lose sight of its value proposition. Within a project, a new feature is wasteful when no one asks for it, no one uses it, or it's requested late. A shadow process or database can pop up when the primary entity doesn't accomplish the expressed purpose. These scenarios involve months of waste among many people.

For an individual project, waste typically originates early in the project. This is because, when governed by a software-centric methodology, upstream assets contain more ambiguity than downstream assets. The Elegance methodology minimizes ambiguity in the upstream assets to minimize waste throughout a project.

Waste exists across projects too. It originates with inadequate listening, expectation setting, alignment, and attention to detail. It results in diverging assumptions, perceptions, and expectations when information swirls around in meetings, emails, and people's heads (tribal knowledge) instead of residing in the right asset (documentation). The Elegance methodology formalizes cross-project rigor to minimize waste.

Iterations and Errors

A mistake repeated more than once is a decision.

~ Paulo Coelho (b. 1947), Brazilian lyricist and novelist

The book's introduction explained five limitations of the Agile methodology and casually mentioned that Agile champions an 'iterative' process. Each iteration repeats a two-week window (called a 'sprint') to build code and deliver features for presentation to the customer. Agile's process is a response to Waterfall's long-lived projects that allegedly have one shot at getting things right.

But 'iterative' is a counterproductive word choice because it hints at tolerating multiple opportunities to get things right. Iterative feels permissive—like saying errors made now can be fixed "in another iteration" without consequence or concern. This is not merely a semantic exploration; it occurs in practice. It's common for the Agile culture to let the team off the hook for not getting something right the first time. But rework can be costly and have significant economic, morale, and reputational damage. Agile over-reacted to a legitimate, but secondary, flaw within Waterfall.

However, critiquing being iterative does not champion irreversibility. A culture of reversibility is better than a culture of irreversibility since it reduces risk. Choices related to process and technology are typically reversible. Choices about people and sales are often irreversible. Be more delicate with decisions about people and sales since their potential for regret and waste is higher.

Another source of waste and rework is errors—the opposite of perfection on the quality scale. There are three types of errors: sampling, systemic, and systematic.

Typical sampling errors include omitting a stakeholder name, mis-sequencing a process flow, or ignoring a metric for a month. Sampling errors are isolated events, and damage is limited.

In comparison, systemic errors are recurring—due to ignorance, neglect, or more urgent matters. Typical systemic errors include producing arbitrarily timed status reports, performing a lessons learned exercise only in a crisis, and letting a roadmap grow stale. These situations repeat and have a pattern. The neglect and waste across numerous teams and projects can compound and grow exponentially.

Systematic errors set themselves apart because they require effort to keep in place. They exist because a stakeholder purposefully supervises the continuation of the errors. As dramatic as it sounds, this is equivalent to corruption or sabotage. Examples in innovation include skipping an annual performance review, prohibiting specific documentation, and hiring outside vendors without a legitimate scoring process.

For a sampling error, the proper response might be, "Be careful!" For a systemic error, the proper response is, "Get it fixed!" And for a systematic error, the proper response is, "Get out!" In every case, detecting a mistake is a cause for celebration, because you avoid tolerating the waste.

Aiming for perfection is counterproductive, but early detection of an error enables early resolution and keeps damage low. Although a policy of continually searching for the next biggest error feels paranoid initially, you can do it in a way that fosters safety, learning, humility, and resilience. Adopting such a policy minimizes waste.

Chaos

The most important role of a leader is to set a clear direction, be transparent about how to get there and to stay the course.

~ Irene Rosenfeld (b. 1953), American business executive

Chaos is wasteful if you embrace peace, progress, and clarity. Chaos is delightful if you embrace mischief, anxiety, firefighting, and drama. Eliminating chaos keeps waste low.

A healthy team works toward agreements in an 'agreement factory.' Chaos-lovers cultivate disagreements and contribute to a 'disagreement factory.' A love of chaos might be accidental or innocent, but people exist who consciously try to generate chaos. Stability makes them uncomfortable; chaos is their comfort zone. They love to start fights, watch fights, and until they achieve power for themselves, they qualify as anarchists.

These people might be employees on paper, but they act in bad faith, and their contribution resembles an infiltrator, not a team member. They don't want to be governed or managed. The waste they can generate is boundless.[5] To put chaos into financial terms, chaos is wasteful because

---

5   Chaos-lovers have one special talent: they excel at divergent thinking. Their favorite question starts with, "What about ...?" When your team exhausts ideas, reaches paralysis, or risks groupthink, consult your chaos-lover. They enjoy playing devil's advocate and thrive at generating ideas for a team that's stuck. But typically, chaos-lovers are terrible for an innovation culture.

it magnifies 'switching cost.' Switching cost is a form of marginal cost. Chaos joins VUCA as a culture trait with high marginal cost, which is why it is financially responsible to combat both.

Chaos also undermines strategic thinking and action. It undermines habits and is the opposite of automation. It forces a team to worry solely about survival, the mechanics of the team, managing a river of surprises, and minimizing occasions of burnout and meltdown. All these are wasteful.

Collaboration and Competition

Competition makes us faster. Collaboration makes us better.

~ Unknown

Innovation and teamwork require both collaboration and competition. Healthy and unhealthy versions exist for both. Choosing healthy versions of each is important because unhealthy versions create tremendous waste.

Unhealthy competition is hurtful and predatory. It takes forms such as cheating in sports, academics, and business. Unhealthy competition erodes into finger-pointing, defamation, and lawsuits.

Unhealthy collaboration is called collusion—a secret partnership to capture money and power. Collusion is attractive to someone who believes they would lose a fair competition and believes the expected benefits of the unethical collaboration outweigh the potential penalty. Businesses might collude on pricing. People collude in cliques.

When playing by the rules, sports, academics, and businesses embody healthy competition. Tradeoffs and tough decisions within a business or for an individual are forms of healthy competition. They bring out the best in people.

Healthy collaboration, meanwhile, equates to good teamwork; and in business, teamwork is everything.

You can be ambitious and goal-driven without engaging in unhealthy, unethical competition or collusion. When you are ambitious in a healthy way, spectators want to watch you, and your competition wants to learn from you. Others want to collaborate with you, invest in you, and invest in your team. When you are hyper-collaborative, others want you to reach your goals.

Healthy collaboration and competition minimize waste. They shape an excellent innovation culture, and they're good for profitability.

Changing the Game

If you don't like the game, change the game.

~ Brett Saraniti (b. 1970), American professor of economics

Innovation work requires a network of individuals, teams, and companies. Goals exist at each of these levels. To motivate constructive behavior at each level, many organizations treat business and teamwork like a game. As the game progresses, some parties benefit and enjoy the game more than others. If a player isn't reaching their goals, they might get frustrated.

The least disruptive response to frustration is for a person to leave the game. For a team, this equates to a person quitting the team. Some team members depart with civility; some depart as sore losers. Quitting might be in the best interests of the team and the individual, but it might mean that skills, knowledge, and relationships leave the team with them. The departing player could join a different team that competes with their former team. But timely quitting minimizes waste for one party—maybe both.

A more disruptive response is for a player to change the game. Now, not every instance of changing the game is bad. In business, a company might reinvent a category. Netflix changed the game of entertainment. Uber changed the game of public transportation. AirBnB changed the game of overnight stays. Changing the game is a form of innovation, and if customers like it, it's not wasteful. But awful forms of changing the game include cheating when a player decides the rules don't apply to them. Cheating—and legal action—are wasteful.

If a player becomes sufficiently frustrated or feels like they are losing their dignity, they might resort to destructive behavior: i.e., destroying the game. It's like throwing a fit, saying, "If I can't get my way, no one can!" In business, examples include stealing intellectual property, embezzlement, sabotage, or physical violence. Destructive behavior is wasteful behavior.

A game is a valuable metaphor for innovation decisions because they have so much in common. Both have winners and losers. Both need governance and incentives that are fair and balanced. The game should last as long as it is fun and productive. Winners and losers need to maintain dignity for themselves and dignity for others. Ideally, players are poised in victory and gracious in defeat. When that stops, the game becomes wasteful.

A game has a great culture when the losers feel the costs are fair and the benefits are sufficiently balanced. A game has a great culture when it minimizes a player's motivation to leave, change, or destroy the game. A game has a great culture when its losers stay in the game.

If you host a team, you are effectively hosting a game. You shape the culture of both. If you don't manage the game on your terms, someone else will change it on their terms.

Matchmaking

The greatest failure in life is being successful in the wrong assignment.
~ Myles Munroe (1954-2014), Bahamian evangelist and minister

Playing matchmaker is a playful and powerful approach to minimizing waste. Innovation resembles matchmaking since it disrupts the status quo away from a poor match (or a lack of a match). Three scenarios that deserve matchmaking are excess supply, excess demand, and mediocre fit. In each instance, addressing upstream issues (raising upfront costs) can dramatically reduce downstream issues (ongoing marginal costs), effectively reducing waste.

Excess Supply

An excess supply equates to too many people for the work in scope. For example, if a city's dining 'factories' (restaurants) have too many chefs, someone must work upstream and generate more opportunities to prepare food successfully and profitably.

The solution for 'upstream' also applies to innovation. If you have an excess supply of innovation workers, you must work upstream and identify more innovation opportunities.

Excess Demand

Excess demand equates to overwhelming problems among customers and stakeholders. The problems might be significant in quantity, complexity, or lifespan. The problems persist because incentives are insufficient to work on the problems. A straightforward solution is to pay the short-staffed chefs and innovation workers more, incentivizing their work. The sooner the right incentives are in place, the less waste you have.

Mediocre Fit

A third scenario is supply meets demand, but the quality of the match is mediocre. It's common to have early enthusiasm—a honeymoon phase—erode into mediocrity. Teamwork stays mediocre when players lack awareness, desire, knowledge, or ability to raise the value of their current assignment; they might even reassign themselves to a more valuable match. To allow a better fit to emerge, you must lower barriers and increase incentives to change. This also requires upstream work.

Reducing waste in your culture is straightforward. An obsession with purpose is a powerful statement to minimize waste; there is no value in ambiguity. Scrutinize your meeting and email etiquette. Forgive sampling errors, minimize repeated systemic errors, and eliminate purposeful systematic errors. Promote healthy competition and collaboration and put a stop to unhealthy versions. Encourage ambition and ruthlessly remove chaos-makers. Place incentives upstream to keep your culture productive and well-matched among contributors and stakeholders.

The tools of Innovation Elegance qualify as 'tough love.' They govern ruthless discipline. But tolerating waste undermines purpose, the employee experience, and profitability. Employees who embrace waste are a bad fit for their organization.

Vigilance

An ounce of prevention is worth a pound of cure.

~ Benjamin Franklin

To keep your innovation factory healthy, your team needs vigilance. The goal of vigilance is to minimize negative surprises. Vigilance also aims to minimize marginal costs in the future by investing in upfront costs. This will maximize the ability of your factory to always maintain your desired speed and quality.

The question isn't *whether* your innovation work will have surprises—the question is *when*. Surprises and blind spots that will jeopardize speed and quality always lurk for your team. Successful vigilance finds the blind spots before they find you.

It is possible to be too vigilant and engage in micromanaging. Excessive vigilance strangles momentum, autonomy, and professional growth. It's also expensive, since it means two or three team members do the work intended for one person. Too much vigilance suggests fundamental problems like distrust and paranoia exist. It's only justified when an error has devastating irreversible damage, such as losing a relationship with an employee or customer.

A lack of vigilance often takes the form of overpromising about time and cost. Because it's normal to want to impress or please others (especially management), it's common to overpromise on a schedule—painting a rosy picture or sharing the most optimistic scenario. Occasions of falling behind are rampant. Some projects procrastinate grasping or acknowledging reality, cloaking status reporting in denial. Overpromising and simplistic optimism ignore vigilance and cause negative surprises.

Phase Containment

A small positive vibration can change the entire cosmos.

~ Amit Ray (b. 1960), Indian author and humanitarian

When vigilance is too low, problems not only appear: they survive and become more significant. Increasing vigilance increases a good form of friction. It slows work down momentarily, but overall, vigilance improves quality. In the immortal words of John F. Kennedy, the 35th President of the United States, "The time to fix your roof is when the sun is shining."

Waterfall methodology formalized vigilance in the concept of phase containment. Phase containment is the conclusion that every defect is one order of magnitude more expensive to fix for every subsequent phase of work it survives. A defect surviving past one phase of work costs ten times the original cost to fix. A defect surviving two phases of work costs 100 times the original cost to fix. Thus, within a project, upstream work poses a higher risk to innovation quality than downstream work.

A modest problem is omitting data fields in a webpage design and discovering it one phase later. A significant problem is omitting a stakeholder in early-project collaboration. A major problem is starting a project whose value proposition is lower than five other projects.

Less literal, but still detrimental, versions of phase containment exist with cross-project work. Neglecting individual mentoring, team-level lessons learned, and change resistance allow poor behavior to magnify negative impacts over weeks and months. In this way, cross-project work poses a higher risk to innovation quality than project-specific work.

A generic solution for phase containment increases vigilance by adding stakeholders to the work. Natural candidates to assign are stakeholders who have recently seen the problems downstream. This transparency for additional stakeholders reduces surprises.

Successful phase containment requires high transparency and low ambiguity. Minimizing ambiguity minimizes wasteful behavior and negative surprises. Phase containment keeps innovation costs and value propositions where you expect them.

Removing Barriers and Bottlenecks

Lead, follow, or get out of the way!

~ Thomas Paine (1737-1809), American philosopher and revolutionary

Successful vigilance decreases and removes bad forms of friction such as barriers and bottlenecks. Removing barriers and easing bottlenecks increases team speed and avoids a work slowdown.

In innovation, a barrier equates to someone unable or unwilling to do their assigned job. They might be on vacation, flooded with too much work, or silently resisting advancing the work. Whatever the scenario, the solution might be the opposite of the solution for phase containment: removing a stakeholder, instead of adding one as in phase containment.

But not all barriers and bottlenecks are equal. Some barriers relate to work where any delay postpones the ultimate project completion date. These barriers—these assets—sit on what's called the 'critical path.' Proactively managing these bottlenecks improves the speed of the factory as a whole. For barriers off the critical path, a modest delay doesn't immediately postpone the ultimate project completion date, so you don't need to reassign the unavailable stakeholder immediately.

You have a big problem if you don't know your project's critical path. The critical path exposes your project's bottlenecks. If you don't know the critical path, you don't know which delays postpone the ultimate project completion date. When you know every bottleneck, you know the exact work to remove an assignment if there is an unacceptable delay.

Bottlenecks never disappear entirely; they just shift. When you fix a bottleneck by changing an assignment or completing work, a new bottleneck becomes relevant. This is just the nature of a factory.

To maximize phase containment, increase heads-down work by adding stakeholders. They are healthy friction and keep quality high. To minimize barriers, decrease heads-down work by removing a stakeholder. Some stakeholders become unhealthy friction and removing them keeps speed high.

The Five Verbs framework makes vigilance easy. The simplicity and transparency of assignments minimize surprises, increase healthy friction, and decrease unhealthy friction.

Variability

Any customer can have a car painted any color that he wants,
so long as it is black.

~ Henry Ford (1863-1947), American automobile pioneer

When innovation teams reflect on what they can improve, especially when a project fails, the typical first response is 'communication.' Communication has practically infinite variables, so labeling a teamwork problem 'communication' is vague and useless. This is one reason the problem has survived for decades. No existing methodologies, management, or leadership books have fixed this communication problem.

Just as the Elegance methodology combats VUCA, it combats variability to address this communication problem. Specific approaches to language and word choices (such as Five Verbs) reduce variability, ambiguity, and improve discipline in communication.

The typical innovation team is all over the place, using exotic words and careless word choice. Plenty of professionals love to impress with an expansive vocabulary. There are times and places for expansive vocabulary, such as fiction writing, song lyrics, and poetry. Project documentation is not that place. Expansive vocabulary takes time to interpret, edit, and agree upon. It's laborious, which means that it's expensive. You can significantly reduce marginal cost in communication by reducing the variability of word choice in project documentation.

The variability in language also shapes your team's culture. High variability promotes uncertainty, complexity, and ambiguity (three parts of VUCA). Low variability promotes focus, certainty, simplicity, and clarity. Governing language variability is a straightforward way to combat VUCA. A skilled innovation team *can* receive and work with highly variable language, but propagating VUCA language makes projects more difficult than they should be. Your stakeholders—current and future—love innovation teams whose outputs are simple and not intimidating.

Adopting these tools represents an upfront cost. Although this cost looks unattractive at first, think about the impact on your marginal cost. If that isn't attractive enough, think about your competitors governing their variability at such a low marginal cost and how to compete with that.

The Elegance methodology is not so narrow that it mandates that cars have a single color. Instead, the methodology prevents cars from having an infinite number of colors—infinite variability. Once Henry Ford governed for color, he could turn his attention to more sophisticated work. Once you govern for language and word choice, you can turn your attention to more sophisticated, customer-centric work.

Active Versus Passive Voice

Words are like keys. If you choose them right, they can open
any heart and shut any mouth.

~ Unknown

One of the most common topics in a business writing course teaches the difference between active voice and passive voice. Active voice includes the subject of the sentence—the actor—directly acting on the object, whereas passive voice focuses on the object, and can omit the actor. Converting passive voice to active voice, "The box is loaded," becomes "Marcus loaded the box."

Passive voice, such as "The customer is sent the new terms," cripples several assets including scripts, process flows, training materials, and status reports. In passive voice, you don't know who or what does the work, unlike when you hear, "The system sends the new terms to the customer."

Passive voice is a pointless form of high variability since an unknown quantity of actors might apply to the phrase. As a form of ambiguity, passive voice hurts speed. The question of who or what doesn't go away; it just forces someone to answer the question in the future. The team is better off clarifying who or what is executing the action from the start rather than procrastinating.

Don't build in ambiguity when you can clearly govern language in documentation. Active voice in your documentation is good language governance. It reduces variability, laboriousness, and marginal cost. Writing in active voice should be automatic. Making it a habit is a form of automation.

Speech Minutiae

It is our responsibility to communicate so clearly that we are understood and so precisely that we cannot possibly be misunderstood.

~ Todd Hunt, American ad executive and business humorist

Word choices often cause misunderstandings within a team. Casual and careless misuse of nouns, verbs, and adjectives might not be devastating, but it can distract and dilute attention from what a writer wants to emphasize. The human brain can typically interpret the writer's meaning and avoid mistakes, but technology is unforgiving. It interprets language literally, so unnecessary variability can create unexpected results.

Nominalization can cause noise when defining and executing processes. Nominalization is using a word as a noun when using as a verb is clearer. Examples include writing verbs such as create, select, and terminate as nouns (e.g., creation, selection, and termination). "Create a picture" is clearer than "the creation of a picture." "Select a song" is clearer than "the selection of a song."

One careless scenario is writing in the future tense when the present tense is clearer for future readers. Examples include process documentation and training materials that explain what a person or a system will do in the future once the project is done. Within a few days of the work, the future tense is not a problem. But months later, the new process is in place, and the language remains in future tense. It can be distracting because the information and instructions are in the present. The documentation doesn't magically change verb tense from future to present. To avoid noisy variability, write in present tense, especially in process documentation and training materials.

Another careless scenario is writing in the past tense. The suffix "-ed" can cause confusion. Using the word onboarded as an adjective for a customer describes the context of a customer. Using onboarded as a verb describes a past action on the customer.

Stating onboarded in a status report to reflect recent work is appropriate. Stating onboarded in process documentation to reflect upstream work or the context of a customer is prone to be presumptuous, redundant, or extraneous information. To minimize confusion in process documentation, write in present tense.

In casual conversation, these scenarios rarely create a problem. But in documentation, where precision matters for interpreting verb tense, nouns, and adjectives, they are ambiguous, confusing, and wasteful. Write in the present tense and check if a noun you're writing is clearer as a verb.

Five Verbs

Repetition is the mother of learning, the father of action,
which makes it the architect of accomplishment.

~ Zig Ziglar

Another form of counterproductive variability is verb sprawl. Many teams experience verb sprawl in a project plan, status report, or task list. Verb sprawl contributes to VUCA, since it's time-consuming to interpret exotic, haphazard word choices. Verb sprawl is clutter that dilutes attention from the most valuable aspects of a project plan.

My noteworthy experience with verb sprawl involved a project manager from an outside consulting firm. He presented his firm's default project plan for the tool we were installing. The plan contained around two hundred tasks (no problem) and forty different verbs (big problem). The verbs included *initialize, determine, consult, set, develop, establish, check, finalize, socialize,* and more. Among the verbs, it was difficult to tell which tasks represented documentation, which represented a meeting, and which represented a simple configuration in the tool. This exotic word choice reflects low discipline and high ambiguity. Verb sprawl like this creates waste in your innovation factory.

Including meetings and emails in your project plan is a slippery slope into micromanaging and infinite variability. Projects require countless meetings and emails. Their dependencies, assignments, and dates are fluid. Labeling them 100% complete in a project plan doesn't represent any significant accomplishment.

A project plan that standardizes a handful of thoughtful verbs is low variability, high discipline, and solves the ubiquitous communication problem. The only work necessary in your project plan is what is worth documenting as a team. You can organize that work into a cadence of exactly <u>Five Verbs</u> (hence the term Five Verbs): draft, review, revise, approve, and distribute. This framework controls variability and minimizes the marginal cost of planning and status reporting. These Five Verbs increase the 'automation' of planning.

Draft

Just get it down on paper, and then we'll see what to do with it.
~ Max Perkins (1884-1947), editor for F. Scott Fitzgerald
and Ernest Hemingway

Following the Elegance methodology, the verb 'draft' is the first station in your agreement factory as it builds an asset portfolio. The drafter is any person suitable to get enough substance on paper for the team to perform the next step; the team understands that the draft is often far from perfect the first time.

The draft step is *the* place for personal brainstorming, divergent thinking, and floating outside the box ideas. Ideal assignments for 'draft' include neutral individuals, excellent facilitators, and junior employees.

Review

You can expect what you inspect.

~ W. Edwards Deming (1900-1993), American engineer, professor, and author

Review is the start of enforced collaboration and the exercising of your collaborative advantage. The review step is the place to leverage the expertise of your tribe—the so-called hive mind. Do note that the review step is one action to optimize globally, not locally. It gives your team license for creativity, divergent thinking, and even for getting a little crazy.

Review is the place to exercise safety, inclusivity, and a sense of belonging because, although not everyone gets their way, everyone gets their say. Ideal assignments for review include non-neutral stakeholders prone to optimizing locally.

Revise

Revision is the heart of writing. Every page I do is done over seven or eight times.

~ Patricia Riley Giff (1935-2021), American author of children's books

The revise step is the collaboration space and the sanctuary of teamwork. The revise step swaps divergent thinking for convergent thinking.

The ensemble has the right and responsibility to narrow options, eliminate alternatives, and make both easy choices and difficult decisions. Through revisions, the team continues to optimize globally.

The revise step hosts the boxing ring of ideas, not a boxing ring of people. It is the place to disagree without demonizing, encourage task conflict, and neutralize personality conflict. Teams wanting a collaborative advantage lean into inclusion and cognitive diversity. Ideal assignments resemble the assignment for draft—sometimes even the same person.

Approve

> *The task of leadership is to create an alignment of strengths so strong that it makes the system's weaknesses irrelevant.*
> ~ Peter Drucker (1909-2005), Austrian-American
> management consultant and author

With great authority comes great accountability. For the people who make the big bucks, Approve is where they make their money. After one or two 'review and revise' windows, the team should feel sufficient alignment. At this point, there is a tipping point from the value of continuing work on the asset to declaring it GETMO. As long as the content is simple and straightforward, assigned individuals approve the document.

When agreement eludes those assigned to review and revise, the team needs someone to play tie-breaker. A person assigned to 'approve' is that person. A good tie-breaker avoids optimizing locally and instead makes decisions that optimize globally (if a tie-breaker is vulnerable to bias, recusal is in order).

Proactively designating a tie-breaker reduces tension and delays. Prior knowledge of a tie-breaker encourages task conflict and neutralizes personality conflict.

Historically, instead of the word 'approve,' some projects used the term 'signoff.' It still appears, but projects use the term less. One likely reason is that it is a poor fit for the reality of innovation work. Even though a team completes an asset, they might learn something later within the project that justifies an unplanned revision to that asset. The term signoff hints at rigidity and finger-pointing. It can suppress having the new information propagate appropriately.

The term 'approve' is less punitive and static. It hints at an inspection and subsequent green light to proceed. Ideal assignments for approve include managers of other contributing stakeholders.

Distribute

> *A lack of transparency results in distrust and a deep sense of insecurity.*
> ~ The 14th Dalai Lama (b. 1935)

Documentation gets a bad reputation when it sits on a shelf. Every asset has some stakeholders who don't contribute but are authorized to know what was agreed upon. The distribute step is the final station in the agreement factory. The last verb instructs someone to ensure the asset is available for additional appropriate stakeholders to see. Ideal assignments resemble assignments for draft—sometimes even the same person.

Managing with Five Verbs converts personality conflict into task conflict. The Five Verbs framework anticipates and shapes the decision process to govern debates. Every big win contains 1001 small wins, and every repetition of Five Verbs is a small win for your team.

Verb Sprawl

> *There is no greater impediment to the advancement of knowledge than the ambiguity of words.*
> ~ Thomas Reid (1710-1796), Scottish philosopher

Anything not covered by Five Verbs is just talk. It's a meeting, a visit to someone's desk, an email, or a long discussion chain. As necessary (and often enjoyable) as all those are in team communication, they are endless; it can become counterproductive to manage them, and their casual results are soon forgotten. Five Verbs ensures that the work is durable.

In a project plan, using verbs outside the Five Verbs framework is not a crime, but they represent work that has low transparency and durability. Shrugging at Five Verbs instantly reduces the value of the work. Adding miscellaneous tasks to a project plan turns the plan into a public dumping ground and impulsive to-do list. It reinvents the wheel and dilutes attention from the all-important collaboration governed by Five Verbs. Five Verbs facilitates communication by reducing remedial communication and replacing it with lasting documentation.

Skeptics of documentation claim it is just more work. In effect, skeptics believe rehashing a pool of forty haphazard verbs is easier to plan and execute than five standardized verbs. Rejecting verb sprawl and adopting Five Verbs is an elegant framework for less work. Months in the future, the output of Five Verbs *is* the work.

A solution for the ubiquitous communication problem must minimize ambiguity in setting expectations and maximize the incentives to meet those expectations. The solution must prevent behavior often seen in teenagers: wanting many of the rights of adulthood without the responsibilities of adulthood.

It's human nature to have some coworkers want the same thing—rights without responsibilities and authority without accountability. A trendy phrase is that team members 'want a seat at the table.' That's understandable, but a seat at the table isn't free. Five Verbs sets the table for rights and authority. Five Verbs bangs the table for responsibility and accountability. Five Verbs guarantees integrity in having a seat at the table.

Organizing your innovation team's work with these five, standardized verbs might seem like petty governance of word choice. Yes, it is ruthless discipline. But the many benefits for the entire employee experience show that controlling variability this way is a small hinge that swings a large door.

The framework right-sizes work. Work that is too small to justify these steps should be bundled with other assets or be euthanized altogether; work that is too big to manage with these steps can and should be partitioned into separate assets.

The simplicity and transparency of Five Verbs creates peer pressure not to be a bottleneck or have too much going on at once. They reduce time and energy on the mechanics of project planning and reporting. Five Verbs combats wasteful communication and fatigue. The verbs govern pacing work with durable quality at a sustainable speed.

If you don't know exactly what to document (assets) or what terminology (verbs) manages them best, the number of tasks to plan, complete, and report becomes infinite. This kind of project—and this level of variability—forces teammates to constantly reinvent the wheel, inevitably creating blind spots for the team. A project with clarity about its assets and the tasks worth managing casts a stable, manageable number of tasks to assign, schedule and track. This kind of project has a confident path for team success.

Instead of a meeting factory or an email factory, your team is an agreement factory, building an asset portfolio that is valuable for months and years. The Five Verbs framework prevents sprawl and enforces simplicity, transparency, and collaboration. These tools to control variability are what you could call, 'culture disguised as language governance.' They are the critical ingredients to resolve the decades-old cliché of project communication problems.

Automation

Remember, technology is a great servant, but a terrible master.

~ Stephen Covey

Over the past decade, the innovation world's emphasis on data has skyrocketed. This obsession has undermined the value of process. Neglecting process governance increases ungoverned processes, manual work, data integrity problems, and chaos.

Data centricity also hinders identifying work that technology should take over from humans—in other words, automation. A data-driven company intends to be pragmatic,[6] but the unintended consequences undermine automation.

When Waterfall methodology was the norm, a project phase called requirements preceded a phase called design. If the project included process-oriented specifications, these specifications appeared in the requirements phase. The next phase, design, included specifications that supported and

---

6   For some organizations, being data-driven originally meant that instead of making decisions with bias, ego, or going with your gut, you operated with pragmatism and attentiveness to stakeholder perspective. This well-intended version of being data-driven is being warped by the version that undermines process and automation.

served those requirements, such as a webpage layout and database structure. Just as design served requirements, data served process expectations.

But Waterfall—then and now—isn't religious about process governance. Agile isn't either. Poor process governance has a devastating domino effect on employees and customers. Information stored in a database is often poorly validated, resulting in polluted and untrusted data. Distrust in a primary database causes employees to build alternate, 'shadow' databases and shadow processes to circumvent the original process and database.

All this shadow work is labor intensive. Investment in process governance has a sizable upfront cost, but avoiding the investment results in high ongoing marginal costs—the precise opposite of agility.

Process Governance

If you can't describe what you are doing as a process,
you don't know what you're doing.
~ W. Edwards Deming

Poor process governance harms more than just data integrity. It hides what the business might try to delegate from humans to technology. In this way, de-emphasizing process governance is the root cause of poor automation.

Solid process assets clarify the circumstances, tasks, and sequence of work assigned to technology actors. Without clarity, database designers won't include database fields to support the processes, and programmers won't build the tasks into the code (and related software components). If tasks and logic aren't in the code, they stay in people's heads with countless assumptions about assignments, alignment, and accountability. If a programmer doesn't put it in code, the work stays manual. If neither data nor code serve the process, automation is not possible.

Because poor process governance is so common, but the spell of data is so strong, organizations incorrectly conclude they have a data integrity problem.

A common response points to the database content, saying, "This information is wrong," and modifying the data. To address repeated data problems, companies often launch projects to establish one "system of record" and a "single source of truth." This obsession with data has normalized data dumps, data lakes, expensive content management systems, and data-centric language like 'a 360-degree view of the customer.'

This fascination with data reflects a serious problem: when you're good with a hammer, everything looks like a nail. For data-driven professionals, every solution is also data-centric. But this data centricity is short-sighted and only serves to 'put lipstick on the pig.' Data doesn't pollute itself. Processes—whether legitimate, shadow, manual, or automated—pollute data. Amid the obsession with data and data integrity problems, the cliché that reflects reality is 'the process is broken.'

Make no mistake: it is more time-consuming and intellectually demanding to fix process than it is to fix data. Fixing a process requires accepting an upfront cost to reduce marginal, ongoing costs. When an error originates in database design or code, fixing it likely requires hours or days. When an error originates in defining process governance, fixing it can take weeks or months.

However, fixing a process problem *as if it's a data problem* doesn't fix the problem; it tolerates and operationalizes data manipulation. Diligent process governance prevents a slow drip of surprise administrative tasks and 'death by a thousand cuts.' It's valuable to get this right the first time.

Understandably but unfortunately, many software vendors sell tools that elevate data over process. Vendors cannot govern their customers' processes, so that is not their business. But vendors do have the incentive to elevate the importance of data over process and champion their expertise in data storage and structure. This undermines their customers' process governance and, in the bigger picture, distracts and discourages customer-centric innovation. The data and the system itself become the master of the customer. The user experience erodes into simplistic data capture and retrieval, while the business logic never works its way into the code. Instead of the business running the business, data runs the business.

This outsources and offloads critical data *decisions* to an outside company, keeps only the manual data *tasks* for yourself, and defends this business arrangement as 'data-driven.' You or such a vendor might romanticize the relationship as 'automation partners.' A more skeptical view is that of a glorified, electronic filing cabinet that imposes a vague or non-existent process onto employees for upkeep. These tools cultivate a higher sense of obligation than vendors want to acknowledge. Humans are at their mercy.

A good process fosters sincere adoption of technology, while a poor process fosters obligation and coercion to use technology. Employees are captive victims of poor processes. Employees might not quit, but their contribution is more laborious and expensive.

Customers are less captive to poor processes; they simply leave, hurting revenue. If a problem exists with data, but the process is mostly acceptable, employees and customers tend to be forgiving. This is because a data problem is likely an isolated event—a sampling error. In contrast, a process problem reveals patterns (systemic errors) or purposeful oversights (systematic errors) that have a more profound origin.

Logistically, getting process governance right—therefore, getting automation right—is not mysterious. Getting it right requires religiously documenting the process,[7] as governed by Five Verbs, before you delve into technology and data-oriented specifications. Process assets force technology and data to be servants of the customer experience. Instead of technology running the business, process governance ensures that business runs the business.

Prioritizing process over data applies to both structured (production) and unstructured (voice of the customer) data. Both kinds of data are critical; however, the processes to capture and store the information are different.

---

7   For example, one way to describe automation pertains to process flows (swimlane diagrams) and the boxes (actions) assigned to human actors or technology actors. Automation is reassigning a box away from a human and assigning it to a technology actor.

Both need process governance to enable automation of the work and avoid runaway manual work and high marginal costs.

Although data doesn't pollute itself, processes do. Ungoverned, processes are prone to erode. Any stakeholder might have the incentive to cut corners of a process at any time. This condition justifies jobs and careers formally titled 'Continuous Process Improvement.' For better and for worse, it's a never-ending job.

Context Management

Content is King, but Context is Queen, and she runs the show.

~ Gary Vaynerchuk (b. 1975), American entrepreneur, author, and speaker

Another form of process neglect is prioritizing content management over context management. Vendors, tools, and projects for content management are common. However, competition in the content economy creates information overload and subsequent waste.

Context management, in comparison, adds the dimension of time. Context is the customer's journey: past, present, and future. It involves empathy and creativity. Instead of a single source of truth, context equates to a 'single script of truth.' Instead of a system of record, context equates to a 'script of record.' Context management governs the entire customer lifecycle and examines any source of waste with suspicion.

Process governance conveys a culture of discipline. Context Management shapes a culture of empathy. When you know the context of customers, you can segment them, apply personas, and meet each customer where they are.

For example, an auto manufacturer might create customer segments for everyday drivers, business drivers, or recreational drivers before tailoring car features. A clothing fabricator might segment gender, geographical location, and age before tailoring the season's line-up.

Without customer context, an employee cannot anticipate customer needs, and a company's products and services are doomed to be one-size-fits-all. Without context, decisions for the customer experience are inevitably spontaneous or even impulsive. They require an experienced, expensive employee that needs to handle multiple contexts. Lacking set context causes high marginal cost.

Rigor for *context* management leads to having the right *content* 'just in time.' It minimizes the waste of time and materials. Context Management cultivates standardization and habits, enables less-experienced employees to serve customers, and leads to low marginal cost. Not only does context management reduce marginal cost, upfront costs to understand the customer are not formidable. Favoring context management over content management facilitates automation.

People like automation because technology reduces cost, labor, and risk, often expressed as the eight D's: work that is dull, dirty, dangerous, difficult, demanding, demeaning, delicate, and dear. Technology serves and improves human experiences. But the elevation of data and technology re-sequences work and inverts who's serving whom. This inversion misuses data and technology, undermines automation, and creates high ongoing, marginal costs.

Automation that takes undesirable work off our hands reminds us that technology is a wonderful servant. Technology and data centricity creating work for humans remind us that they are horrible masters.

A healthy innovation culture commits to the discipline in process governance to minimize polluted data, waste, and cost. It commits to empathy in context management in order to meet the customer where they are, anticipate their needs, and thus reduce labor. Process governance and context management nurture automation and improve the human experience.

Ease

Do what is easy and your life will be hard. Do what is hard and your life will become easy.

~ Les Brown (b. 1945), American politician and motivational speaker

So much in business and life feels difficult. It feels like work. We stand in awe of people who make it look easy while performing at the top of their game. However, behind the polish and highlight reels, the elite in any profession have discipline, dedication, and an extraordinary work ethic.

Change leaders have a financial incentive to make innovation easy. That's why leaders install methodologies—to make innovation easier and more profitable than if no methodology were in place.

The Elegance methodology strives for ease through a framework called earn versus burn. This chapter educates on this term, distinguishes healthy and unhealthy earn and burn, and gives details so you can balance activities for optimal productivity and profit.

Earn Versus Burn

If you don't have the time to do it right, when will you have time to do it over?

~ John Wooden (1910-2010), American basketball coach

Earn versus burn is a framework to distinguish what the customer values from the labor and resources needed to complete the valuable work. Documentation is evidence of progress as an innovation team 'earns' its way toward durable value in an asset portfolio. Meanwhile, a team 'burns' time and money in energy-consuming meetings and emails to build and maintain their asset portfolio. Meetings and emails are evidence of activity, but not proof of productivity.

If asked, "How is the project going?" you might respond in terms of earn or in terms of burn. Examples of earn include, "We've agreed on the roadmap!" "We've agreed on the project charter!" or "We've agreed on our target metrics!" A response in terms of burn can be humorous, but also sounds silly and bitter. Examples include, "We've exchanged 1000 emails on the project," or "One month of meetings so far!"

A simple analogy for earn versus burn includes different ways you could report progress on a road trip. An earn report says, "We covered 900 miles and have 100 miles to go!" A burn report says, "We drove twelve hours and spent $300 on fuel!"

Another example of earn versus burn is analogous to monthly revenue per square foot. Consider a small factory (1,000 sq. ft) with revenues less than a large factory (10,000 sq. ft.). That's no surprise, but for an apples-to-apples comparison, calculate revenue per square foot. Now consider how well your team creates value with a small footprint. High value with a small footprint is 'doing more with less.' A team with a high earn versus burn ratio makes it look easy.

Thoughtlessly or obnoxiously minimizing burn, or even maximizing earn, is simplistic and dangerous. Doing this on a road trip risks safety and a speeding ticket. Doing this with a team risks careless mistakes and burnout. A sweet spot is an approximate balance among your asset portfolio, meetings, and emails for a sustainable intensity of earn versus burn.

Making Meetings Work

The magic to a great meeting is all of the work that's done beforehand.
~ Bill Russell (1934-2022), American professional basketball player

Among complaints related to teamwork, meetings rank high. Many organizations have a love-hate relationship with meetings, yet they are often a default activity for teamwork.

Meetings are professionals' comfort zone and addiction. In attempting to feel included and to feel like they are producing value, many professionals get caught in meeting gridlock. Common complaints include, "We have too many meetings!" and "I can only get real work done after-hours!" Meetings are considered the business world's biggest waste of time because there are dozens of ways to execute meetings poorly.

Meetings can't and shouldn't disappear. They are intrinsic to teamwork, especially for complex and nuanced topics. A meeting conveys that the topic is complex enough to justify synchronous communication, as email or voicemail (asynchronous communication) are only effective for collaboration on simple topics. But meeting governance is important. Specific behaviors at different stages of meeting execution ensure your team maximizes the long-term value of meetings and minimizes fatigue and resentment. These behaviors make meetings look easy and expedite genuine progress on assets.

Conventional thinking says every meeting requires an agenda, but even the best-informed and constructed agenda lacks long-term value when it dilutes attention from the actual collaboration space. Months and years in the future, your team won't reference the agenda; your future team will want to see the result of the collaboration. Of course, that collaboration space is an asset, timed and synchronized among other assets in your agreement factory.

Meetings are the lubrication that keeps the Five Verbs aspect of your agreement factory moving. The first and last verbs—draft and distribute—are never the meeting's agenda,[8] but the middle verbs—review, revise, and approve—are. This everlasting agenda is a simple, unambiguous way to ensure that every meeting serves the asset portfolio and keeps the factory moving. This everlasting agenda ruthlessly renders the question, "What is the purpose of this meeting?" obsolete.

Creativity should appear *inside* the asset, but creativity in setting the agenda is just noise. Anything outside the Five Verbs framework suggests the topic is not worth documenting or collaborating on, implying that the topic isn't valuable for the team's future.

Before the meeting, monitor that the invitations and RSVPs are appropriate. Some meetings need divergent thinking (ideation and brainstorming), some meetings need convergent thinking (narrowing options and making decisions), and some meetings need both. Attendees should know what combination to expect.

Everyone should also know beforehand who the approvers for the asset are. Approvers also serve as tie-breakers when consensus eludes the team.

At the start of every great meeting, a couple of things are in place:

1. Everyone is prepared for the conversation: i.e., every attendee has read the asset and prepared reactions, questions, and suggestion.

2. Everyone is reasonably punctual to avoid a significant disruption. One person's last-minute conflict should not derail work or penalize others.

Throughout, a valuable meeting has several signs of strong discipline:

1. Attendees are attentive and stay within the topic/scope of the asset. The group logs off-topic items (barriers, risks, issues, and questions) separate from the asset.

---

8 An exception is when a team chooses to distribute sensitive or celebratory content in a meeting instead of a repository or email.

2. The conversation contains the right mix of divergent and convergent thinking.

3. Attendees work hard for a rigorous and complete asset. They work hard to uncover inconsistencies, waste, and blind spots.

4. Ideally, edits are easy enough to make within the meeting but, if not, recording the conversation for playback and revisions is acceptable.

In the spirit of 'seek first to understand then be understood,' senior members allow and encourage junior members to carry the meeting as much as possible. Senior employees ask questions, express caution, and explain acceptance or rejection of an idea in a spirit of encouragement and positive reinforcement. They curb anyone, especially themselves, from rambling or monopolizing the conversation.

A healthy, valuable meeting has several signs of empathy. The meeting is not hurried, but attentive and focused. Everyone feels included, heard, and safe; no one feels suppressed, and disagreements are aired without demonizing. Ideas are diverse; bad ideas are handled with grace to minimize embarrassment. Attendees undermine personality conflict, explore authentic task conflict, and seek positive surprises.

Silence is comfortable and welcomes the next comment. When comfortable space exists between comments, it shows that interruptions are discouraged and that it's important to let every speaker finish their statement. Interruptions are few, polite, and even create humor. Ideally, attendees have the attentiveness and courage to sense acquiescence and silent dissent and invite this potentially awkward information to be shared.

The end of a healthy meeting contains several traits:

1. Everyone is satisfied that they exhausted their contributions. Everyone gets their say, even if they don't get their way.

2. Everyone is at peace that the tie-breaker chooses among options if consensus eludes everyone else.

3. Agreement takes a back seat to alignment, i.e., a common under-standing of the results and decisions.

4. Conversational rigor reinforces that the meeting was a good use of time and the proper collaboration forum.

Attendees should consider and agree on the value and urgency of any additional meetings, or they should approve the asset. In either case, the facilitator verifies that appropriate stakeholders have access to the new version of the asset soon after the meeting.

Reserve meeting placeholders well in advance, possibly at regular intervals, such as weekly, to maximize schedule visibility and minimize conflicts. It's not a crime to reschedule a meeting, but it *is* a grave discourtesy to stakeholders to procrastinate scheduling an important meeting.

Effective Emails

Writing a long email is easy. Writing a short one is hard.

~ Shane Parrish (b. 1976?), Canadian author and entrepreneur

Email has been, and continues to be, a fantastic productivity tool for business. Email reduces the need for business travel, meetings, and even phone calls. It's an incredibly efficient method of sharing information.

Email is the right channel for information that requires a few minutes to read and that loses its importance after a few weeks. It's the right way to share information with low sensitivity, emotion, or risk. Email responsiveness is always a good thing, although an email suggests the topic is less time-sensitive than a conversation or meeting.

The ease of email has a few downsides and abuses. First is the avalanche of email in the business world. It's easy to overcommunicate by abusing email. In the clutter of email overload, it's easy to overlook an email or an important messenger. Repeated 'reply alls' indicate the topic is more complex than what email can resolve.

Email is a perfect channel for sharing information, but it's a poor collaboration space. Email is also the wrong repository for information that the team wants to reference months and years into the future. As soon as an email includes information that belongs in a collaboration space—an asset—abort the email and instead add the information to the right asset. And if you see an email with a CYA tone, it is a warning sign that the team has systemic alignment and accountability problems.

As valuable and necessary as meetings and email are, organizations cannot meet and email their way to success. Their durability is low, and their laboriousness is high. As such, they contribute to fatigue.

Mindful companies monitor the balance of the three communication channels: assets, meetings, and email. Most companies should dramatically increase their emphasis on assets and decrease their addiction to meetings and email. Periodically (perhaps as part of monthly lessons learned exercises), poll stakeholders about their perceptions about the balance among meetings, email, and documentation. Rebalancing toward documentation increases the upfront cost of collaboration but lowers its marginal cost. Meetings and email are wonderful servants, but horrible masters. Aim instead for steady (not lumpy) progress in your agreement factory's production of an asset portfolio.

Documentation Makes Life Easier

I hate writing. I love having written.

~ Frank Norris (1870-1902), American journalist

What makes innovation easy over the long term is documentation that exists inside a shared repository. Documentation is a good 'master' for teamwork; meetings and emails are wonderful 'servants.'

That said, teams shouldn't create documentation for the sake of documentation. They should create documentation because it fosters simplicity and transparency. Documentation promotes humility, accountability, and other desirable culture traits. Documentation reshapes a culture to ease teamwork and behave with more empathy.

Another way documentation shapes your organization's culture and discipline is by modifying your team's cost profile. A low documentation team has low upfront costs but a high marginal cost. A documentation-oriented team has high upfront cost and low marginal cost. This is another example of discipline being unpopular at first but compelling once the hard part is over.

Not all documentation is equal or even valuable. The last two decades of innovation have normalized several documents[9] that don't reduce the marginal cost of teamwork or stay valuable beyond a few days. These documents are counterproductive because they harbor ambiguity and dilute attention away from the assets that reduce marginal cost.

Another misuse of documentation is ignoring or avoiding Five Verbs. It's common to draft a document (often at the request of a superior) without assignments for the other verbs. If a document lacks priority for team members to review, revise, or approve in the short term, it might never be valuable to the team. Collaboration for your asset portfolio requires Five Verbs.

Governing with any other verbs is not collaboration. That work is waste at best, or chaos at worst. Orphaned drafts make teamwork difficult. Simple, transparent collaboration requires documentation and governance by Five Verbs.

A few documents have the goal of *listening* instead of *collaborating*. Employees draft and review these only—instead of executing all five verbs. These documents are one person's view of the world that their coworkers either accept or reject. This is documentation that enables disagreeing with-

---

9 Examples of common but counterproductive documents include meeting minutes, data strategy, and technology roadmaps.

out demonizing. They help to manage expectations and perceptions about team health and behavior.

Innovation teams are happier, more productive, and more valuable when they combat the addiction to 'burn' (meetings and email) and instead become addicted to 'earn' (the proper assets). Emphasis on the right assets makes work easier today and for your team in the future. A stable asset portfolio reduces each individual project's footprint and makes innovation as a whole easier.

There are several benefits to a stable and standardized asset portfolio. Standardizing your asset portfolio prevents *under*-communication; no one can claim ignorance and skip something when everyone knows what innovation teams need to document. A stable portfolio also undermines *over*-communication and discourages employees from thoughtlessly or impulsively dreaming up a new asset.

The asset portfolio is not eternally static. The business world will push you to innovate how you innovate, and you will find additions and variations to your asset portfolio. In whatever way innovation teamwork evolves, a stable asset portfolio right-sizes a sequence of agreements and minimizes negative surprises.

Another dimension to ease forming the asset portfolio is synchronization. Synchronization is knowing which assets must be sequential and which can progress concurrently. A synchronized schedule has high visibility for an entire project. Such a schedule minimizes spikes, droughts, and surprises.

These three characteristics of an asset portfolio ease staffing decisions and help you play matchmaker between work supply and demand. Systematic visibility and synchronization avoid reinventing the wheel and dramatically reduce labor related to the mechanics of project management. This freedom increases opportunities to find innovations and stakeholders upstream from in-flight innovation work.

Healthy ease in innovation boils down to two tasks related to earn versus burn. The first task is to execute and synchronize Five Verbs to build a stable, standardized asset portfolio. The second task is synchronizing meetings and email to keep the Five Verbs process moving while avoiding meeting gridlock and email overload.

Overcoming addictions to meetings and email might not be easy, but once you are addicted to a harmonious collaboration on an asset portfolio, innovation teamwork becomes easy.

Autonomy

This is my dance space. This is your dance space.
~Johnny Castle, *Dirty Dancing* (film, 1987)

Your autonomy is defined by your identity and your boundaries. Your autonomy is a product of your self-sufficiency. Don't be casual in your decisions about your innovation team's autonomy; too much or too little autonomy is high risk and has far-reaching effects.

Autonomy involves centralizing and decentralizing decisions and activities. Responsibility and authority can be concentrated or dispersed. Workers and organizations can be coupled or independent.

Several terms speak to autonomy, and you can use these words in a positive or negative light. Low autonomy in a positive light is 'partnered,' and in a negative light, 'micro-managed.' High autonomy in a positive light is 'self-sufficient' or 'empowered,' and in a negative light, 'abandoned,' or a 'single point of failure.'

The best place in innovation teamwork is a happy medium that avoids risky extremes. Your risk tolerance and perception of what is risky inform your decisions that shape your autonomy.

Too Little Autonomy

Control leads to compliance. Autonomy leads to engagement.

~ Daniel Pink (b. 1964), American best-selling author

At an interpersonal level, leaders and followers sometimes operate with low autonomy. A manager might micro-manage an employee. A team member might want or need a lot of attention from colleagues. Extremely low autonomy undermines independent thinking, professional growth, and the benefits of teamwork and cognitive diversity.

At an organizational level, too little autonomy results when a company delegates too much to outside companies. Over the past fifty years, large organizations in high-wage countries have commonly outsourced specific work (typically its most mundane operations) to organizations in low-wage countries. The rationale for outsourcing is to reduce cost. The outsourcing company might see the work as less attractive, less core, and less strategic for the company's mission. For many reasons, the outsourcing company wants to avoid paying high labor rates in its own country, even if outsourcing increases risk.

Another common form of outsourcing is when an organization hires outside specialists (i.e., consultants) to do the core work of a large project. Consultants are expensive. Companies pay a premium for consultants' temporary services instead of relying on internal employees to execute all the functions of innovation work.

You hire outsiders when you believe their contribution is better for long-term profit when compared to relying exclusively on internal employees. When you do so, you accept the increased dependence on resources that you do not directly control. Ideally, the outsiders minimize surprises related to quality, and the perceived risk goes down.

However, countless organizations outsource some of their most strategic and sensitive innovation and operations to people who give their attention to more than just their primary client. Too little autonomy increases risk.

Too Much Autonomy

I think this society suffers so much from too much freedom, too many rights that allow people to be irresponsible.

~ Boyd Rice (b. 1956), American composer, artist, author, and painter

The other extreme also poses risks. A worker with too much autonomy is not collaborating. A lone wolf is effectively not part of a team. Work done in a personal silo is likely work that no one else values.

An organization has too much autonomy when it consults too little, yet lacks the skills to execute independently. Too much autonomy results in people and organizations having 'enough rope to hang themselves.' Retaining or experiencing a blind spot are signs of too much autonomy.

All humans value a certain amount of agency. We dislike being highly dependent on another person, technology, or organization (like a vendor or consultant). Overcompensating to gain independence can cause having (or being) a single point of failure (SPOF). Factories, systems, and networks are vulnerable when they possess SPOFs.

This SPOF concept applies to people, too. Backups and options are rarely free, but often a worthwhile insurance policy to mitigate disaster.

While autonomy and self-sufficiency are great, good leaders don't abdicate accountability or abandon their team's work. Good leaders stay close enough to work to provide discipline, safety, and harmony with the bigger picture. Good leaders set up their team for success, and that requires avoiding too much autonomy. Too much autonomy increases risk.

"Just Right" Autonomy

Interdependence is a choice that only independent people can make.

~ Stephen Covey

Before the twenty-first century, a company often looked outside itself to plan and execute innovation, viewing it as a peripheral secondary activity. But innovation has become a primary activity. Individuals, companies, and even countries are experiencing more intense change. Whether innovation is on a company's own terms or another's terms, innovation is now mainstream. As a result, many organizations are less inclined to outsource innovation work and the associated skill building to consultants.

Skill continuity (as opposed to resource continuity) for innovation is more valuable as innovation matures as a core competency. Skill continuity reduces the organizational friction to invest, invent, and innovate. Organizations that do this are responsive and ready to pivot. Organizations own these resources, so workers don't have other clients, and they have fewer diverging agendas. From the lens of an organization, the long-term trend is toward more autonomy.

Savvy organizations build innovation skills in-house as *two* core competencies: *identifying* the NMV (next most valuable) work (i.e., project ideas) and *completing* the NMV work (i.e., the projects themselves). The Elegance methodology aims to reduce your in-house cost to *complete* the NMV innovation so you can spend more on *identifying* new NMV innovation.

For individual members of a team, autonomy is also a healthy state. Being empowered to make and act on decisions is motivating. An innovation factory that is autonomous and able to decentralize decisions and labor is faster and more predictable. It has less idle work and fewer idle workers. Healthy autonomy includes flexibility, transparency, and trusted handoffs.

The right level of autonomy avoids high cost and high risk. Healthy decisions and a culture of autonomy keep your stakeholders safe and their work sustainable. The happy medium is when all parties first feel upbeat about their own identity, then enthusiastically help others to succeed. In the immortal words of flight attendants everywhere, "Be sure to secure your oxygen mask first, then help others."

Elasticity

Blessed are the flexible for they will not allow themselves
to become bent out of shape.

~ Robert Ludlum (1927-2001), American author of thriller novels

S tability and confident expectations are desirable for an innovation factory. But gradual and sudden changes are inevitable. Skilled change leaders are ready to handle disruption to their peaceful equilibrium.

A healthy innovation factory has a certain rhythm. At equilibrium, it has a certain speed, tension, and familiarity. At equilibrium, the motion is safe and sustainable. Levels of franticness or idleness are low, and negative surprises are few.

But nothing is static. Disruptions happen. Team activity and productivity can gradually become imbalanced or unsustainable. If a team is rigid or unresponsive, small disruptions can have a disproportionate impact on team performance. Alert change leaders are responsive to small changes and adjust team behavior to restore safety and balance. Wise adjustments are a type of elasticity.

Major disruptions occur, too. Positive surprises arrive through opportunities with new team members, customers, and a larger circle of stakeholders.

Negative surprises arrive in the form of departing team members, departing customers, and new competition.

A culture of elasticity encourages a team to respond to new information and achieve a new equilibrium. A new equilibrium involves a new set of team members, customers, and stakeholders. It involves new vulnerabilities and alternatives. A healthy new equilibrium has balance, options, and low idleness.

Idle Work (Demand)

If you're at zero backlog, it's a whole lot easier to change priorities.

~ David Allen (b. 1945), American productivity consultant

Idle work has other names like 'productivity at a standstill,' 'neglected opportunity,' and 'value and profit waiting for us.' Idle work appears in operations (neglected customers and stakeholders) and in innovation (neglected improvements).

It's understandable to put idle work in a positive light; after all, how can too many customers be bad? Whether at equilibrium or in a disrupted state, business leaders might consider too much demand—idle work—a good thing. But idle work implies a company is slow-to-market. Slowness in the market hurts revenue. Lost revenue is easy to spot in an operational context (e.g., customers seeing a queue and walking out the door), but it is less visible in an innovation context (e.g., the delay of a new product or service).

Idle work is often mislabeled. Operational concepts such as Six Sigma and Lean Manufacturing have brought value to many organizations and misuse to others. Some organizations call themselves 'lean' when they are simply understaffed. Having too few employees to serve customers' needs translates to complacency in idle innovation and lost revenue.

Backlogs are common at an organization's equilibrium, and a market disruption magnifies awareness of it. Throwing more people at the problem is a common, but simplistic, first response to feeling short-handed. Consider two other responses before doing so.

First, decentralize the work by spreading out assignments among more existing team members. Then, organize and synchronize the pace of the smaller tasks. This reduces the workload for over-assigned and unavailable team members.

Modest and steady reductions in one person's workload ease a bottleneck and can dramatically improve team performance. Easing bottlenecks can be as straightforward as excluding overly taxed team members from meeting invitations, emails, and assignments for some assets. It's counterproductive and impossible to get idle work perfect, but tuning assignments periodically at project bottlenecks creates a new equilibrium of idle work.

Once you fix work distribution and synchronization, avoid a short-handed or 'lean' team. You will want to pursue your valuable backlog of serving customers and innovating for them, and you will feel confident increasing the size of your team.

A low amount of idle work is not a problem; in fact, it's a healthy sign and a scenario for managers not to intervene. But high, persistent amounts of idle work cripple your culture's elasticity and justify manager intervention. When new information meets high idle work, the team is short-handed. A manager's only option is reassigning workers away from valuable in-flight work to compensate. Even when new information is a positive surprise, changing expectations for employees and customers can have adverse effects. Keep the level of your idle work modest. This is a sign of elasticity.

Idle Worker (Supply)

That the devil finds work for idle hands to do is probably true. But there is a profound difference between leisure and idleness.

~ Henry Ford

The inverse of idle work is an idle worker. Your culture and equilibrium for idle workers impact profitability less directly than idle work, but they shape your team's elasticity.

As hinted in the previous section, it's common to misrepresent being short-handed or understaffed as being 'lean.' Some managers have a strong distaste for even a casual sighting of a relaxed worker, concluding they are disengaged or lazy. These managers see zero idleness in their *workers* as zero idle *work*, which is precisely wrong. These managers often have no consequences for overburdening their teams, even in cases of burnout and turnover. Some managers even resent workers taking time away for vacation, training, and medical care.

A persistently high level of idle workers is also mismanagement. It shows poor synchronization, poor training, and even marginalization of workers. Doing this at your equilibrium neglects and ignores workers. At its worst, it holds workers captive and abuses their time and money.

Instead of always being heads-down in their work, it's healthy for workers to have some idle time to be heads-up. This gives workers the space to think strategically, explore, and learn. A modest level of idleness encourages workers to work refreshed and recommitted.

Elasticity requires a modest level of idleness among workers. Modest idleness results from healthy process governance. There is an optimal amount of slack in any process, and it is greater than zero. This provides time for workers to have healthy breaks. Breaks are preventative maintenance and keep stakeholders injury-free.

If you compare equal amounts of idle work and idle workers, idle work has a more significant impact on a business's revenue. A one-day delay in a project (idle work) might impact many customers or employees. A worker being away from their job for one day (idle worker) rarely impedes an entire project or numerous stakeholders.

Working toward perfection in idleness is counterproductive. The effort for small amounts of improvement is high. The cost of perfection almost always outweighs the benefits. Modest levels of idleness are good enough—and optimal—for an elastic team.

Keeping levels of idle work and idle workers modest is nothing short of matchmaking. Keeping levels of idle work modest is customer-centricity. Keeping levels of idle workers modest is employee-centricity. Succeeding at both shows attentiveness, responsiveness, resilience—or in a word, elasticity.

Balanced Operational Pipeline

Everything in moderation, including moderation.

~ Oscar Wilde (1854-1900), Irish writer and poet

Your organization is a factory whose operations might produce pizzas, patients, mortgages, or baseball games. Your factory's inputs and outputs have equilibrium levels, and your factory rhythm is vulnerable to disruptions. Balance plus alternatives equals elasticity. Both are important to handle disruptions, which will minimize damage from negative surprises and maximize opportunities from positive ones.

An operational imbalance is a scenario that you have been overproducing or underproducing. For example, if your factory produces pizzas, you make too many (or too few) pizzas than you can sell in a day. Balance is the scenario that you produce approximately the ideal quantity of pizzas for your customers. That ideal quantity is when the revenue and the cost of your last pizza sold are equal: i.e., marginal revenue equals marginal cost, and profit is zero.

If you experience a disruption, say from customers or competitors, this sends the demand for your pizzas up or down. This, in turn, impacts what you charge for those last pizzas—your marginal revenue. Disrupting a balanced equilibrium is not a big deal, since the disrupted marginal profit is small. This is elasticity.

Contrast that with a disruption to an imbalanced pipeline. If you were overproducing or underproducing, marginal profit might be far from zero. Disruptions have an immediate and significant impact on marginal profit, and change in demand becomes a big deal. It can expose missing out on even more revenue or exacerbate unprofitable production. This is poor elasticity.

You could similarly analyze a disruption originating from your suppliers, which sends the supply for your factory input up or down. This disruption impacts your marginal cost. Disrupting a balanced equilibrium is not a big deal, since disrupted marginal profit is small. Aim for an approximately balanced equilibrium and install supplier alternatives to minimize the downside of disruptions and surprises.

Precision is not critical; a marginal profit of approximately zero is healthy. Balance and alternatives are the keys to elasticity in your operational pipeline.

Balanced Innovation Pipeline

I'm as proud of many of the things we haven't done as the things
we have done. Innovation is saying no to a thousand things.

~ Steve Jobs (1955-2011), American pioneer
of the personal computer revolution

Your innovation team is also a factory—producing assets that define customer and employee experiences. Your team's inputs and outputs have equilibrium levels, and your project rhythm is vulnerable to disruptions.

An innovation pipeline has a balanced equilibrium when you synchronize all in-flight projects and future projects to optimal profitability. What matters are the actual projects, their sequence, and their intensity. If an organization adds or accelerates projects while balanced, it over-innovates and hurts profitability. If an organization cancels or decelerates projects while balanced, it is under-innovating and hurting profitability.

A balanced equilibrium should feel steady, not lumpy. Any lumpiness in innovation rhythm is only justified by a disruption that originates from customers, stakeholders, or competitors. Disruptions to a balanced innovation pipeline are no big deal, because the impact on marginal profitability is small. Disruption to an already-imbalanced innovation pipeline likely exacerbates conditions of both over innovating and under innovating.

An elastic innovation pipeline activates all projects that positively impact long-term profitability. Disruptions to such a pipeline are not a big deal, since paused projects have a low return on investment (ROI). Your least profitable projects are fantastic options to cancel or pause.

A team is ready for disruption when it knows which projects to pause and which projects to start if new information makes them more or less compelling.

Imbalances in an innovation pipeline are common. Most companies under-innovate. They have more prospective profitable projects than actual projects. This imbalance hurts elasticity, causes companies to miss out on customer value, and reduces their profit.

Over-innovation exists in a couple of different forms. One form is the gradual erosion of expected ROI. Some projects' attractiveness fades, especially as new prospects appear.

The second form is projects with a sketchy origin. Pet projects, favoritism, and kingdom-building are ego-centric and not customer-centric. Both situations justify suspending a project.

Contrary to conventional thinking, suspending a project is not a terrible event. It could be a sign of a few positive things:

1. Suspension shows that the team is aware of the project's low ROI.

2. It shows that the company is humble, courageous, and responsive enough to do something about it.

3. The competing project might be a positive surprise and an excellent opportunity for the company.

Elastic companies that emphasize assets (and do not rely on meetings and email) resume projects more easily when one returns to being among the most compelling projects.

In prioritizing and sequencing projects, precision can be counterproductive since value propositions are bound to shift. What is imperative is that an innovation team thoughtfully completes a Business Case and Project Charter for every project. That rigor is always enough for a team to avoid huge mistakes and start the most compelling projects in (approximately) the right order, size, and intensity. Avoid hasty, thoughtless starts to projects, avoid analysis paralysis, and proceed with a thoughtful, approximately optimal, answer.

To maximize long-term profit, your innovation pipeline must be elastic. Elasticity requires a balanced equilibrium and possible alternatives in the face of disruption. With elasticity, you are ready for positive and negative surprises.

Elasticity for Employers and Employees

It's easy to make good decisions when there are no bad options.

~ Robert Half (1919-2001), employment pioneer

Every project has an end, and every employee/employer relationship has an end. Embracing this and planning for it helps to convert these ending relationships from a negative surprise into a positive transition.

Gone are the days of twenty- and thirty-year relationships between employers and employees. There's a term for the opposite extreme—the Gig Economy.[10] The timelier topic is whether ending the relationship is a surprise and whether all parties can transition with a positive tone.

---

10 The Gig Economy is a labor market characterized by the prevalence of short-term contracts or freelance work as opposed to permanent jobs. The term gig borrows from the performing arts community, where comedians, musicians, and others are paid for individual appearances known as gigs.

Look to the stock market for a model to manage relationships. Every investor wants to buy low, sell high, and avoid the opposite. Savvy employers and employees aim for the same. Leaving on poor terms equates to 'selling low.' It takes work, discipline, and maturity to know when to 'sell high' in the stock market, and the job market requires the same. But exiting the employment relationship while everyone's stock is high preserves value and options long after the formal relationship ends.

Barriers to joining and leaving employment relationships are dropping. Some resist this evolution, preferring job security and partnerships that last long beyond their unique value. For some, being trapped in a relationship that has exhausted its mutual benefits is a comfort zone.

Elasticity avoids the trap of a comfort zone. It rejects exclusivity and embraces inclusivity. Elasticity is recognizing that all parties have great alternatives and not feeling threatened by others' attractive options. A culture of elasticity makes transitions less of a surprise and more like a student's graduation.

In our younger days, everyone moved on every four years, and in the spirit of growth, innovation, and disruption, we should embrace that in our working relationships. This mindset converts negative surprises into positive transitions. This elastic mindset helps business relationships to 'buy low, sell high.'

Blaming the Hiring Decision

If you get the culture right, most of the other stuff will just take care of itself.
~ Tony Hsieh (1973-2020), American internet entrepreneur
and venture capitalist

Many companies have employees that 'don't work out.' People in power decide that someone is performing poorly and terminate their employment. The company concludes the problem was followership, not leadership. The expectation is that followers need to meet leaders where they are.

If this event occurs once, the company treats it like a sampling error. People in power take minimal responsibility and blame the error on the hiring decision. If the event occurs multiple times, the company might treat the scenario as a systemic error, take more responsibility, and change its recruiting or interview process.

Companies with the most humility give followers the benefit of the doubt and conclude, "We set them up for failure." Only the humblest organization comes to this conclusion that the problem is not followership but leadership.

Leadership is among the most common topics in business books. This is because the world has leadership problems. If the world felt it had followership problems, bookstores and talk shows would feature those authors. No one should be surprised that the world has leadership problems, because good leadership is more difficult than followership.

If you hire an intelligent, experienced person and they're doing their best in their followership role, their job should not be difficult. People setting others up for failure is what makes innovation difficult, and thus a person's manager has the most influence on a job's difficulty. It stands to reason, then, that leadership is a bigger problem than hiring decisions.

Instead of practicing leadership, people in positions of power often practice 'climbership.' Climbers are competitors, not collaborators. Climbers behave as followers, not as leaders. A common scenario is for a climber to see a talented junior employee (sometimes reporting to the climber) as a threat and want to get rid of them. Actively or passively, the manager sets up a talented employee for failure, possibly characterizing the event as a hiring mistake.

Hiring mistakes are expensive. Hiring mistakes of leaders are even more so. But the most expensive of all are methodology mistakes. The success of too many companies is dependent on specific leaders and followers. The way to be less vulnerable to people (and perceived hiring mistakes) is to have a methodology that is elastic towards people and their competitive human nature, which means drawing out their collaborative qualities.

Good methodology mitigates performance shortcomings of leaders and followers through teamwork, collaboration, discipline, and empathy. Bad methodology is difficult to rescue, even by a great leader. Good methodology can overcome even a poor leader.

Teams should also expect less that followers meet leaders where they are and instead expect leaders to be elastic, embracing that leadership is not one-size-fits-all for their followers. Leaders need the tools and incentives that set up all stakeholders for success.

The Elegance methodology exercises elasticity by anticipating and embracing diversity in a team so that leaders can meet their followers where they are. Innovation Elegance is a systematic collaboration that doesn't believe in hiring mistakes. It trades competitive advantage for collaborative advantage and trades climbers for leaders.

Elasticity is a valuable culture trait for your innovation factory. Elasticity is not just personalities that avoid being too rigid or too loose. An elastic culture keeps idle work and idle worker levels modest. It maintains balanced operational and innovation pipelines. It converts the end of the employer/employee relationship from feeling like a break-up into feeling like a graduation. And it rejects the notion of followership problems by executing a methodology that sets up everyone for success.

Conclusion: Innovation Culture Like a Factory

Success is the sum of small efforts – repeated day in and day out.
~ Robert Collier (1885-1950), English lawyer, politician, and judge

Your organization is a factory, and your innovation team should resemble one. Managing innovation like a factory instills the discipline you need in your culture to serve customers, thrive with employees, and maximize profit.

Conversations about culture are valuable and important, and the factory metaphor reduces the ambiguity in those conversations. Your factory cares about every dimension covered in this section of the book: economics, speed, quality, waste, vigilance, variability, automation, ease, autonomy, and elasticity. Traits of your culture translate to traits of a factory. The impact is not ambiguous; these traits will impact your bottom line.

Plenty of organizations surrender to conventional thinking about VUCA and accept software-centric methodologies. But messiness is a cop-out, and software isn't what makes innovation difficult. The metaphor of a factory and building an asset portfolio by assigning Five Verbs is people-centric, ruthlessly simple, and courageously transparent. These frameworks will lead your team to innovate with discipline and elegance.

GRACEFUL EMPATHY: CULTURE DISGUISED AS THE ARTS

Empathy is the engine that powers all the best in us.
It is what civilizes us.

~ Meryl Streep (b. 1949), American Actress

In the summer of 2012, a friend invited me to a Chicago restaurant called Nacional 27. It was half-price appetizer night, and they also offered a free beginner Salsa dance lesson. We had a great time. Little did I know that evening was the start of a fantastic chapter of my life: immersing myself in the Latin Dance community in Chicago.

Like most students, I started slowly. But it's not fun to be a beginner on the dance floor, so I had the incentive to learn quickly. Salsa led to Mambo. Mambo led to Cha-Cha and then Bachata. The beginner Bachata team needed another guy, so I joined, and six months later found myself on a more advanced team. Two years later, I was teaching beginner classes to friends, church groups, and at a few bars in my neighborhood.

I learned a ton—from how to twirl a partner to the three Rs of resilience, recovery, and resetting. But most important was that everything I learned *on* the dance floor also applied *off* the dance floor: whether at work or in my relationships.

On the dance floor, I need mechanics and style. Off the dance floor? Yep and yep! On the dance floor, I need to both lead and listen. Off the dance floor? You bet. Even things you might not associate with dancing, like self-esteem and humility, the dance floor reinforced. Superficially, the language of dance was fun and persuasive. But at a deeper level, the *culture* of dance shaped the cultures of my teams in the workplace.

This prompted me to reflect on whether other non-work activities involved elements that transferred to innovation work. Harmony, balance, synchronization … these musical terms applied. Being on-script, off-script, layers of moving parts, and a story … theater concepts applied, too. Improvisation skills were relevant: my team had occasions when the most constructive conversations were about exploration, brainstorming, and sharing small ideas aloud to build upon one another.

This cross-pollination from the world of the performing arts to the business world isn't new. The business world already makes plenty of casual references to the performing arts. You frequently hear, "Are we singing from the same script?" "It can be a delicate dance with this customer," and

"Improvise! Make it up as you go along." But what I wanted to explore was closer to immersion—not merely borrowing the language of the arts, but also their habits, community, and culture.

Reflection on this cross-pollination made me wonder if the parallels were limited to the performing arts. What about martial arts—or even the 'art' of parenting? No one would classify them as performances, but these arenas of intense human interaction had traits and attitudes to teach innovation teams.

Parenting teaches safety, mentoring, authenticity, and self-sufficiency. Innovation teams already casually use related language like 'family' and 'set the table,' but they didn't explore all the parallels. Innovation teams regularly deal with internal conflict, but they lack a methodology for resolution between opponents. Terms like passive-aggressive show that the business world already has language to describe the tension they experience. But the martial art of Aikido teaches you how to neutralize an opponent without destroying them. Innovation teams can learn a thing or two from Aikido.

Parenting and martial arts have another similarity to the performing arts: they focus on other people. A parent's attention is on their children. A martial artist's attention is on their opponents. This attention is a form of empathy. Parents care what their child thinks, says, and does. Martial artists care what their opponent thinks, says, and does. There is an art to this empathy, and it's teachable. Innovation teams can benefit from these arts that are so infused with empathy.

Of course, conversations about empathy can have as much ambiguity as conversations about culture. Reflecting on these six 'empathetic arts' and employing their language, habits, and culture reduce this ambiguity and improve your team's empathy.

Casually, teams already use terms from the empathetic arts. This section deliberately connects arts and innovation teams and minimizes ambiguity to shape a culture of empathy.

The metaphor of the empathetic arts is compatible, and even complementary, with the metaphor of a factory. The factory encourages discipline, focus, and ruthlessness. The arts promote playfulness, growth, and grace. The factory feels cold and harsh. The arts feel warm and inviting. Innovation teams can benefit from this yin and yang, and to sustain a balanced culture, they need it. Every team needs a combination of discipline and empathy. To maximize profit, you want to be intense in both.

Intense discipline resembles a ruthless factory. Intense empathy resembles a graceful factory. An organization with the best discipline and empathy wins for themselves, their employees, and their customers.

Comparing a Graceful Factory with a Ruthless Factory

Creativity is the combination of two or more things that
were previously thought to be mostly unrelated.

~ Unknown

The first section of this book established that conversations about culture don't have to be ambiguous. Dozens of decisions that make up your culture translate to your innovation factory's speed, quality, and so forth. These culture traits unambiguously impact your bottom line.

Yet a methodology solely focused on discipline that uses a factory as a metaphor feels sterile. An innovation methodology needs more than that. It needs to inspire creativity, passion, and purpose. The metaphor of the empathetic arts gives a team license to explore the options with style and grace.

The metaphor of the empathetic arts also minimizes ambiguity. These arts have tangible elements that innovation teams should emulate. The empathetic arts are relatable; everyone has some exposure to them, and many have taken part in lessons or performing. Most of us don't pursue the arts professionally, but value learning the basics and doing the fundamentals well. We might even playfully imagine ourselves as a literal rockstar, movie star, or prima ballerina because working in the arts appears more glamorous than working in a factory.

The empathetic arts and a disciplined factory have similarities, such as an audience, a sense of motion, and tangible outputs. The two metaphors have differences, too: the arts focus on empathy, exploration, and positive surprises, which are foreign concepts to a strictly ruthless factory.

Like a factory, the empathetic arts involve projects that produce tangible outputs and a sense of accomplishment. Every project creates enjoyment for artists and audiences again and again. Collections of different performances form 'asset portfolios' that are financially, socially, and emotionally rewarding. The low marginal cost of maintaining such asset portfolios is attractive in financial and non-financial ways. The marginal effort to re-experience music or a movie can be as easy as pushing play.

Like a factory, the empathetic arts also offer a sense of closure and moving on. Every role and every project comes to an end. Artists and the audience move on to new teams and performances.

The empathetic arts also enforce collaboration—a necessity in innovation. As a performer, you do your best for your team and every new audience. Even if you're a solo performer, you collaborate with those offstage who provide behind-the-scenes value. All of them want to contribute to a wider goal.

The empathetic arts are fun, but they are not all fun and games, and definitely not all fluff. There is plenty of hard work and even discipline. The arts require dedication, because sometimes you dislike your role so much that you want to quit. Fatigue and burnout are common. You stay only to be a steward to the organization, learn what you can, and leave things better than the way you found them. Working toward goals in the empathetic arts requires pacing yourself, pacing the team, and hitting interim milestones—all features of innovation work too.

The success rate of the empathetic arts provides a positive model for innovation. Innovation's disappointing success rate shows up in countless books, events, and online forums, where thought leaders propose how to innovate better, smarter, and more profitably. Innovation's lack of success is also reflected in the increased mental health needs of innovation employees. Large, high-profile organizations conduct frequent surveys about project success and failure to pinpoint root causes—each survey's pie chart more sophisticated than the last.

The empathetic arts have sufficient success to not need thought leaders, mental health attention, or pie charts. The high success rates and world-class talent found in Hollywood, Bollywood, and Broadway receive a ton of attention and praise. But you don't have to be world-class to succeed. The pros had to start somewhere, and local talent is everywhere. The arts world does something right that the innovation world can learn from.

This second section of *Innovation Elegance* explains how the empathetic arts inject collaboration, growth, and grace into your team culture. These culture traits have far-reaching effects on team members, audiences, and your extended population of stakeholders. They give you a collaborative advantage to your innovation teamwork. The skills and culture of the arts make you a more skilled change leader and make your team more valuable.

Common Experiences across the Empathetic Arts

Empathy is the ultimate form of customer insight.

~ Don Peppers (b. 1950), American author and professor of marketing

Culture traits and experiences across the empathetic arts are diverse. Before exploring specific traits unique to each art, this section explains the progression of experiences and culture that are universal throughout the arts. These experiences show how the arts require performers to be thoughtful about where to collaborate versus compete.

For example, artists behave *collaboratively* when they join a team, rehearse in a safe space, genuinely trust, and respectfully listen to each other. Embracing the habits of collaboration in the arts is intense work, but these arts require collaboration. Without collaboration, parenting, symphonies, and dances fall apart. Emulating this intense collaboration in innovation teamwork is a collaborative advantage.

Artists behave *competitively* when they balance their contributions against others, recover from rejection and failure with poise, and embark on new work when one piece of it is good enough. Competition in the arts is intense, and artists' unique experiences cultivate a healthy competitive nature. Emulating this intense competition in innovation teamwork is a competitive advantage.

In general, innovation professionals are not managed to approach collaboration and competition in this same thoughtful way, but when they are, project success rates are high, resembling success rates in the arts.

Everything artists do—collaboratively and competitively—must keep the audience in mind. Financially, the audience 'pays the bills.' Emotionally, they're why artists do what they do. Audiences share positive and negative reports about the performance and keep the artist in business (figuratively and literally). Audiences can be picky or forgiving. They can be intimate or distant. Their relationship with the artists might be long or short. Artists must be who their audience needs them to be.

Just like artists are audience-centric, innovation teams must be customer-centric. Customers keep innovators in business financially and emotionally, can also be picky or forgiving, and can be an intimate relationship of trusted partners or purely transactional at a distance. Successful innovators are who their customers need them to be.

The empathetic arts reward thoughtfulness in these experiences—thoughtfulness about when to collaborate and when to compete, striving to delight audiences. This thoughtfulness creates a culture of empathy, balances intense discipline, and results in high success rates in the arts.

The arts are a model for what healthy collaboration and competition look like. Software-centric methodologies are less thoughtful on this matter. The Agile Manifesto declares, "Working software is the primary measure of progress." The Elegance methodology declares, "A working team is the primary measure of progress."

Artists' Collaborative Advantage

Alone we can do so little; together we can do so much.

~ Helen Keller (1880-1968), American author,
disability rights advocate, and lecturer

Collaboration is being ambitious on behalf of your team. Competition is being ambitious on behalf of yourself. The empathetic arts clarify when to compete and when to collaborate. The arts make both a more pleasant experience for performers.

In the arts and innovation, collaboration appears in:

- Joining a team

- Safety and authenticity

- Trust

- Listening

- Practice

- Rehearsal

In these activities, performers prioritize others' needs. This raises everyone's value and morale and gives the team a collaborative advantage. The inverse is also true—demoting others' needs lowers the team's value and puts it at a collaborative disadvantage.

Conventional heroes and individual superstars are attractive and exciting, but they carry risks. If their behavior damages a teammate's experience, that behavior reduces others' value and the value of the team. In the short term, that behavior can hurt speed, quality, and *customer* experience. Long-term, talented team members seeking a collaborative experience leave that team, searching for a better experience. The best collaborators make their teammates better—a different kind of hero and superstar.

Much of the world is obsessed with being increasingly competitive—and with pursuing a competitive advantage. If you want to distinguish yourself from that, consider how to uniquely collaborate and establish a collaborative advantage. Collaboration has such a large impact on overall results that treating it casually is a big mistake. Good parents, symphonies, and dance partners insist on collaboration to give performers a positive experience. The best innovation leaders insist that their team pursue such a collaborative advantage.

Joining a Team

Great things in business are never done by one person.
They're done by a team of people.
~ Steve Jobs

Like in innovation work, every performer typically joins a team—an ensemble. Even when a single person is on stage, many people help them practice and prepare before the performance. During the performance, people backstage manage props, technology, and the audience. Obvious or not, every performer is part of a team.

To join most ensembles in the arts, everyone passes an audition. To join an ensemble in innovation, everyone passes an interview. As long as the leader polices aggressive behavior, what was competition becomes collaboration.

The arts choose the word ensemble instead of team to convey the meaning in French: together. For anyone to succeed, everyone needs to succeed—together. And for everyone to succeed, everyone must be engaged. A member feeling self-conscious, disengaged, or excluded hurts the culture of togetherness. Innovation and the arts are team sports—collaborative sports.

Most innovation teams don't yet have the same commitment to 'together.' Aggressive behavior among colleagues occurs 'post-audition' more often than any Human Resources department wants its employees to know. Companies are increasingly aware of employee discord and exclusion, and recent trends are positive for improving engagement and inclusion.

Teams in the arts and innovation have diversity in experience, containing beginners, intermediates, and experts. Explicitly or subtly, everyone learns from and substitutes for each other. Beginners hardly do anything on their own; they need constant attention, guidance, and feedback, which brings out empathy in the more experienced team members. Although being a beginner fosters humility, beginners typically are self-absorbed, which can limit their contribution. They are sponges, and often learn by imitation and repetition. They bring a fresh perspective and thus lead in their own way.

A beginner's self-esteem rises, falls, and rises again. Everyone is a beginner compared to someone else, so if your organization is renewing for the future and has sustainability in mind, it has beginners.

Just like teamwork fosters humility, it also fosters vulnerability. Everyone on the team is vulnerable—to the audience and to other performers in the form of opinions and criticism. Vulnerability is a healthy behavior when it's on your own terms—being approachable and coachable by team members and their well-intentioned feedback. There are, however, forms of unhealthy feedback, such as being overly dependent on others for approval and being gullible around individuals with malicious intent. And when a person cannot be vulnerable, they create blind spots for themselves, which ironically amplifies their vulnerability. Having thick skin is healthy. Having a thick skull is not. Vulnerability across a team requires that everyone learns from everyone else.

Many innovation professionals have a colleague whose arrogance reduces the amount of sharing of information across the team. This person's lack of personal vulnerability creates vulnerability for their entire team.

All humans are vulnerable to making mistakes. Mistakes can matter, but damage is often limited, and the individual and the team can usually move on. To quote Miles Davis, one of the most celebrated jazz musicians of the twentieth century, "When you hit a wrong note, it's the next note that makes it good or bad." Imperfections can be a source of embarrassment, but also humor. Bloopers and outtakes sometimes take on a life of their own, revealing that even pros make mistakes and can make fun of themselves. Being able to laugh at yourself is healthy.

A team has countless benefits for its customers, and being part of a team has countless benefits for individuals. Relationships, inclusivity, and a sense of belonging are good for emotional health. At its best, teamwork is a fun, energizing, experience rich with positive surprises. Team members look out for each other to minimize negative surprises. Customers benefit from teamwork because what they buy benefits from multiple specialties across the team. Relationships survive and thrive after the performance and after the project.

Performers—in the arts and innovation—weave in and out of being individual contributors and leaders, alternating between being their best selves and bringing out the best in others. At an audition and as an individual contributor, a performer is competitive, being ambitious for themselves. After the audition and serving as a leader, the best performers are collaborative, acting ambitiously on behalf of the entire ensemble. This sort of weaving is necessary to have a healthy and high-performing team.

Whether in the arts or innovation, working in isolation might be therapeutic for a while, but it doesn't benefit anyone else. Accomplishing anything requires that you join and embrace a team. An ensemble is a collaborative advantage.

Safety and Authenticity

The privilege of a lifetime is to become who you truly are.

~ Carl Jung (1875-1961), Swiss psychiatrist and
founder of Analytical Psychology

To keep their head in the game, innovation professionals need to feel safe performing their job. To do their best work, they need to behave with authenticity. The empathetic arts provide models for both safety and authenticity. This section couples these topics since their absence has a common root cause: a hostile person in power. Hostile people create a toxic environment, deny safety and authenticity, and devalue others' identity and work. A leader who prohibits a toxic environment and enforces safety and authenticity gains a collaborative advantage.

Distinct kinds of safety exist across the arts. Martial arts emphasize physical safety since the actor is being physically attacked. Parenting adds emotional safety since children are naturally emotionally vulnerable. Improv provides a kind of intellectual safety—that whatever instinctively comes to mind and comes out of someone's mouth is, by definition, always okay. Dancing requires social safety. Dancers feeling unsafe with a partner should walk away when they can seize an opportunity.

The safety that applies to innovation professionals is most often called psychological safety. Companies have financial and reputational reasons to provide a safe environment for their workers. An unsafe environment causes employees to shift their attention away from their jobs toward the threat they perceive. An employee who feels unsafe is likely looking for another job, taking their skills and knowledge with them. And former employees are likely to talk about the toxic work environment to their network, hindering a company's ability to attract employees. Safety is a prerequisite for healthy collaboration—in the arts and in innovation.

Authenticity goes one step further than safety. Safety is a predecessor to authenticity because to act authentically, you must first feel safe. Parents exercise safety with tough choices about permissiveness. Children exercise authenticity when facing choices that risk their parent's disapproval. For example, a child might secretly resent piano lessons; a teenager might suppress their sexual orientation; and a young adult might try a career pressured by their parents. Inauthenticity might win for a while, but authenticity is the healthy long-term path.

The art of improv promotes authenticity because players are conditioned to give instinctive responses in the exercises. Each improviser knows that everyone else also has a license for authentic approval. A pause means the player is being too cautious or calculating. Players know that every contribution—surprising or not—is immediately met with an approving, "Yes!" A pause or a hint of inauthenticity is the only response that prompts disapproval—a playful groan from other players.

It's valuable when employees bring their authentic selves to an innovation team. An authentic employee exhibits passion, curiosity, and transparency. They fearlessly raise concerns and new ideas. Honesty and candor without retaliation—combined with civility and poise—accelerate a company to its frontiers of performance and customer-centricity.

A culture of authenticity prevents negative surprises, and fearlessness can generate positive ones. A positive surprise often originates from cross-application. Examples include experiencing a seat warmer in a car, food service to your seat at a baseball game, and a flight attendant doubling as a stand-up comedian. A diverse workforce that authentically shares ideas capitalizes on that diversity by opening new doors and exploring those options.

Feeling unsafe and inauthentic puts a person under duress. Duress elevates the temptation to drop best practices and work in fear. Fear may push someone to run harder, but it will never push anyone to run smarter; fear kills your organization's creative advantage.

Insincerity diminishes value and can even be dangerous. When employees are strictly 'going through the motions,' their best ideas stay suppressed and dormant. They build resentment, exercise silent dissent, and behave with vicious compliance. If an inauthentic employee snaps, employees feel blindsided and betrayed.

The subtle beauty of a safe and authentic environment is how it enhances the speed at which ideas are shared and decisions are made. In such a team, fresh ideas instinctively come out, and there is no incentive to be insincere. Authentic colleagues don't read each other's minds, but they grow to know what their colleagues will contribute; this allows them to anticipate points of view and behave with empathy toward diverging ideas. This culture makes your organization a uniquely attractive workplace—a collaborative advantage.

Trust

The best way to find out if you can trust somebody is to trust them.

~ Ernest Hemingway (1899-1961), American novelist and journalist

Trust is a foundational element of teamwork. With trust, people work together to create and accomplish great things. Without it, people isolate themselves and contrive reasons to compete—a collaborative disadvantage. In the arts and innovation, trust allows individuals to focus on what they do best. Trust includes confidence that others will do their part to serve audiences, customers, and stakeholders. It's a collaborative advantage to have employees this ambitious on behalf of others.

As important as trust is, the word itself is ambiguous. To optimize trust and diagnose mistrust, consider four flavors of trust: competence, integrity, dependability, and benevolence.

1. **Competence.** Competence is a matter of intellectually being able to do the job. Common ways to assess competence include title, education, experience, and observation.

2. **Integrity.** Someone's integrity is good if they set expectations in good faith and report on events to not mislead. A person of integrity 'means well.'

3. **Dependability.** Dependability is someone's ability to execute without surprises. Some people overpromise and overcommit, hurting their dependability. Dependable people avoid negative surprises and follow through on the expectations they set for others.

4. **Benevolence.** A benevolent person knows what is in your best interests, doesn't desire to undermine them, and often wants what's best for you.

Trust exists across the arts in all its four forms. Musicians expect *competence* in other ensemble members to perform their part. Parents hope their teenagers have the *integrity* to not consume alcohol or commit crimes. Actors expect *dependability* from others to show up to rehearsal. Dancers expect *benevolence* in the desire for a pleasant experience from their dance partners.

Healthy innovation teams have all four forms of trust—competence to fulfill their role, integrity in past and future contributions, dependability to execute, and benevolence that everyone wants success for everyone else. Catchy riffs in the industry include the 'say/do ratio' (say what you do and do what you say) and the 'say/want ratio' (say what you want and want what you say). Trust improves with predictable patterns of behavior, minimal negative surprises, and the occasional positive surprise.

Occasional failures—sampling errors—don't damage trust much, but systemic errors do. And systematic, purposeful errors equate to outright malevolence. Conventional wisdom remains true: trust takes time to build, seconds to break, and forever to repair. Low trust prompts time-consuming inspections, increases workload, and slows work down. Trustworthy colleagues accept accountability, possibly apologize, and resume work with humility and vulnerability. Working with transparency makes it easier to trust again. A culture to gain and regain trust is a collaborative advantage.

Trust is valuable because it minimizes waste. Distrust isolates people, cripples teamwork, and wastes time and talent. Trust within a team puts people at ease and promotes confidence, focus, and approachability.

Listening

Leaders who don't listen will eventually be surrounded
by people who have nothing to say.

~ Andy Stanley (b. 1958), Georgia-based pastor

Innovation teamwork is often short on listening. Poor listening hurts learning, alignment, and morale. Good listening improves transparency, safety, and inclusivity. The arts show how essential and valuable listening is. Listening skills are far from universal, so teams that listen have a collaborative advantage.

Business people are wired to be responders, but not to be generous, empathetic listeners. Past business leadership styles involved a lot of talking and directing. For people in power, listening to others seemed optional. But a leader in the arts who fails to listen fails fast. For example, any music ensemble's stated goal is to *make* music. However, the ability to make music is completely reliant on listening to your fellow musicians. Everyone listens to each other to harmonize and identify opportunities for improvement.

Most listening problems in the business world are one-way. When a junior employee is a poor listener, the poor listening ends when they are coached, or they simply lose their job. When a senior employee is a poor listener, that poor listening can persist for a long time. This simple power structure causes listening problems to be leadership problems that last a long time.

Thus, the biggest listening problems are among leaders—people in power. No one wants to receive bad news, but everyone knows at least one leader who, in fact, *refuses* to hear bad news. But when a leader fails to listen, they fail to understand, creating blind spots for their team. Instead of solving problems, leaders who don't listen *create* problems.

Conventional listening could be called 'receiving signals' with our ears. Actors receive visual signals from other actors and theater directors. Dancers receive signals through a collaborative sense of touch. Martial artists receive signals through a combative sense of touch. Whether through our sense of hearing, touch, or sight, receiving signals—listening—is vital to collaboration.

Not all listening is good listening. Poor listening is just being polite and barely quiet enough while waiting for a chance to respond. Good listening commits to understanding and completely receiving another person's message on their terms and timing.

Good leaders listen to what people say and often hear what is unsaid as a person's body language can betray what they are thinking and feeling, and thus visual listening (observing) is crucial too. Great leaders are phenomenal listeners and observers. And if any leader wants to have the last word, they must speak last.

Average listening is often good enough, but superior listening takes collaboration to special places. Children listen to their parents to stay out of trouble. Good parents are great listeners so they can detect trouble and keep their children out of it. Good parents listen patiently and keep the right to speak last. Observing subtleties and attention to detail leads to high quality in all the arts.

Poor listening guarantees poor team decisions. Poor decisions all have a common root: they were made in a vacuum where the decision maker failed to listen. Listening is a muscle—it needs to be exercised to see improvement. Just because people can get away with being bad listeners does not mean there is no cost. The cost of bad listening is that you build things no one asked for and no one wanted.

It is important to note that the visibility of listening differs from the visibility of making noise. Committed listeners risk being perceived as passive (as my friend Lisa observed, when you rearrange the letters of *listen,* you get *silent*).

Every good listener picks their battles—their interventions. Whether you are a music conductor or a change leader, one of your primary goals is to listen to the sound as the ensemble proceeds through the music. You might detect twenty things you would like to coach but choose to share just a fraction of them with the group at one time. You might choose to stop the ensemble frequently, or you might wait until the song is finished before sharing your coaching points.

When the ensemble is not making music, staging a scene, or moving in sync, you also listen to the performers' comments. You are not the only person who might notice something that can be improved. Ideally, the performers under your leadership feel safe sharing their observations and ideas, acknowledging their imperfections, and making their needs known.

That's not to say everything a leader hears is true, accurate, or fair; discernment matters in leadership. It is vital to get a reasonable number of perspectives and understand the biases and motivations among stakeholders. Leaders balance their convictions with openness to other perspectives. A leader listens to the team's input, factors it into their ideas, and then decisively sets the direction in a respectful manner. For example, a choir director might hear varied opinions about the mix and sequence of songs for a concert. The director carefully considers the different ideas and decides what they feel is best for the performers and the audience.

Change Leaders often conduct unstructured listening with their stakeholders, meeting them where they are. This can be time-consuming. At some point, diligent stakeholders will want the leader's response. Like a conductor or director, innovation leaders convert the lack of structure into clear actions. Good leaders minimize instances where team members feel they aren't heard or that their concerns go nowhere.

Leaders work hard to give everyone their *say* and temper expectations that not everyone gets their *way*. Leaders translate less-than-professional remarks into constructive, professional language so their ideas have somewhere to go.

Empathy is a prerequisite to innovation, and empathy requires listening. Leaders are successfully empathic when every team member feels they have a voice and are being heard. If you can't listen, you can't innovate. If you can't listen, you can't lead. The arts exercise a culture of listening.

One of the best-kept secrets in the arts is that the performers have the best seats in the house. It's one thing to listen to performers from the front row balcony; it's another to experience the growth and performance of an ensemble standing on the stage surrounded by your fellow performers. Collaborative innovation teams adopt this lens on listening. You and your team members have the advantage of having the best seats in the house, listening to the collaboration evolve.

Practice

Practice puts brains in your muscles.

~ Samuel Snead (1912-2002), American professional golfer

Just because you were invited to join an ensemble doesn't mean you can stop practicing alone. Martial artists practice on their own or with a video. A dancer practices steps so they're better for their partners. An actor runs lines on their own not during precious rehearsal time with others in the scene.

While rehearsal is an integration of the ensemble, practicing improves the quality of your slice of the material. Diligent performers come to rehearsal prepared. They set their ensemble up for success, while unprepared performers set their ensemble up for failure. Practice contributes to a collaborative advantage.

Similarly, innovation professionals have solo work to prepare for integrated work with their team. Mapping solo work to Five Verbs, the drafter of an asset takes the asset as far as possible. Then they send it to reviewers with enough time before a review session to prepare feedback. Diligent reviewers and approvers prepare their input before a review meeting. This sets up the meeting for success.

The lack of solo preparation is a significant problem for innovation teams. The most common explanation is a heavy workload, so one solution is to reduce the workload for team members who are frequently unprepared. Another solution is to slow the project's tempo to allow more review time for everyone.

Solo, heads-down attention to an asset encourages deep thinking. Generosity with your time frequently leads to the best ideas and breakthroughs, just in time to present to the team. The generosity of personal practice is a collaborative advantage for the team.

Rehearsal

You need to make mistakes in rehearsal because that's how you find out what works and what doesn't.

~ Clarke Peters (b. 1952), American-British actor, writer, and director

Rehearsals are occasions for individuals to collectively align on the material. Ensembles must regularly align to avoid significant divergence. Busy parents need to have dinner with their busy kids to achieve their version of alignment. Innovation teams align on project assets and typically accomplish this in meetings. Rehearsal is an alignment session that contributes to a team's collaborative advantage.

In the arts, the director sets expectations for rehearsals. Ensemble members must adhere to these goals and their sequence. It's important to stay attentive, focused, and responsive to what everyone else does and needs. Improv ensembles show good rehearsal technique by repeatedly setting up team members for success. Improv is the art that prohibits 'watch your back' culture and enforces 'got your back' culture. Everyone knows and executes their role.

The goals and scope of rehearsal typically are the most difficult parts of the material. Team members handle the simple material outside precious rehearsal time by practicing on their own.

For innovation teams, rehearsals equate to meetings. Healthy innovation teams follow every culture trait of rehearsal from the empathetic arts: know your job, do your job, and do *only* your job. Be who the ensemble needs you to be and have your team members' backs. Try to handle simple matters outside precious meeting time and expect that meetings handle the most challenging material.

Since meeting gridlock is such a problem in innovation, it's worth reflecting on how the arts avoid a parallel problem. Gridlock is a case of having too much going on. With their success rate as evidence, skilled art directors prevent having too much going on. They set boundaries by keeping their ensemble's scope modest. The work has a simple, straightforward organization that easily accommodates changes and delays.

Rehearsals detect underperformers. Some ensemble members neglect their preparation or disrupt others. This hurts the quality of the ensemble. A good leader asks what we might call 'freeloaders' and 'deadweight' to leave the ensemble.

Likewise, review meetings detect underperformers on an innovation team. Some underperformers overcompensate by saying too much and monopolizing meetings. Other underperformers are unprepared, unusually quiet, or working in fear. And a third type prides itself on chaos and playing devil's advocate. Underperformers should be coached to change their behavior, or the leader should ask them to bow out of the project.

Rehearsals also show the exciting progress of the ensemble. Progress shows itself gradually as small wins that build in layers upon the last. Musicians layer their melody, harmony, and dynamics. Dancers first practice a routine without music and then the routine *with* music. Actors layer their lines, scenes, and choreography.

Innovation teams similarly galvanize their progress in meetings. Teams layer alignments for a new customer process, technology design, build, testing, and training.

In the arts, a last rehearsal is often called a dress rehearsal. Some innovation teams already use the same term, whereas others employ terms like testing, beta release, or pilot. A dress rehearsal occurs on the performance stage, includes a safe audience, and rarely stops for anything that goes wrong.

In an innovation dress rehearsal, the project team creates an environment for customers to execute the new processes with all the aspects of post-release. But the atmosphere is still technically a safe forum if something goes wrong. Testing verifies that human actors and technology actors fulfill their roles as expected. Just like a dress rehearsal for a concert or a play, the team tries to uncover last-minute oversights and surprises.

For an ensemble, every big win involves 1001 smaller wins, and those appear during rehearsal. For an innovation team, every big win involves 1001 smaller wins, and those appear during review and approval meetings. Every stage has the potential to be exciting and rewarding and, when that's the case, the ensemble enjoys the journey. Rehearsing provides the arena for alignment, small wins, and long-term collaborative advantage.

Bring Out Their Best

*Love is simple: it's two people doing their best to
bring out the best in each other.*

~ Unknown

In the arts, leaders and performers bring out the best in each other. A music director wants to satisfy other performers. Dancers want to make a good impression to be invited to dance again. Good parents bring out the best in their children academically, emotionally, and socially.

Innovation professionals are at their most valuable in a culture that brings out their best. Innovation teams are excellent environments for positive peer pressure and mutual accountability. Day-to-day, employees are focused on their customers. The long-term result is that exposure to diverse teamwork builds well-rounded skills.

The culture of the arts encourages and enforces collaboration. Emulating the arts gives innovation teams a collaborative advantage.

Artists' Competitive Advantage

It is teamwork that remains the ultimate competitive advantage
because it is so powerful and so rare.

~ Patrick Lencioni (b. 1965), author, speaker, and management consultant

Always being *collaborative* and never being *competitive* is naïve. People are ambitious and competitive. That's a good thing, since these improve our quality of life.

In the arts and innovation, competition exists in the following culture traits:

- Balance

- Vigilance

- Rejection

- Failure

- Resilience

In these activities, a performer doesn't prioritize others' needs, because they have to worry about their own. Performers must improve their skills to stay relevant. As a performer increases their value, they gain a competitive advantage.

There is such a thing as being too competitive. Being hypercompetitive is predatory. Predatory actors justify audits and regulatory compliance. Accommodating predators is being complicit with bad behavior. This hurts the customer experience *and* the employee experience; predators destroy value for everyone they touch. Innovation professionals are wise to avoid predators. Isolation is an excellent incentive to stop predatory behavior.

In the arts and innovation, being personally competitive is necessary. Being competitive conveys competence and relevance. It's a personal competitive advantage to show you can work with others and collaborate within a team yet be outwardly competitive for your organization to survive and thrive.

Being thoughtful about your competitive and collaborative advantages—and distinguishing each—is another skill. The Elegance methodology is mindful of being competitive, but it doesn't emphasize competition. Instead, it emphasizes collaboration while accommodating and governing the ambitions of individual stakeholders. The combination of intense ambition and collaboration creates a formidable advantage.

Balance

The team in motion is never in balance, but always correcting for imbalance.

~ Gregory Bateson (1904-1980), English anthropologist
and social scientist

When there is a sense of competition among team members, balance is an advantage. Balance matters to both artists and innovation teams. An imbalance creates resentment among team members, hurting morale and value. Balance boosts morale and a sense of fairness.

Parents face non-stop decisions on balancing what they give their children regarding money, time, and attention. Improv teams aim for balanced contributions across all performers. Musicians strive for balance across all voices and instruments.

It's easy to see what imbalance looks like. Favoritism among children hurts social and emotional health. Imbalance in a choir or a dance partnership is painful to hear and awkward to watch.

The catchy phrase for imbalance is feast or famine. Feasting is a scenario of overdoing something and has a root cause in insecurity. Famine is a case of underdoing something. Its root cause is neglect.

In an imbalanced innovation team, feast and famine exist in two places—thinking and doing. A feast of thought monopolizes ideas, opinions, and decision making. The famine of thought is a team member suppressing their ideas and abandoning their logic. The feast of doing—a heavy workload—causes a hero mentality, fatigue, and burnout. A famine—an uneven workload across a team—causes a victim mentality, passive behavior, and freeloading. These imbalanced scenarios hurt speed, quality, and morale of an innovation team.

A healthy balance minimizes feast and famine. Balance enables an organization to optimize globally. A healthy innovation methodology anticipates potential insecurity and neglect among stakeholders, then proactively manages incentives and assignments to minimize both and achieve balance. Balance in competition is healthy and an advantage for a team.

Vigilance

Beware of rashness, but with energy and sleepless vigilance
go forward and give us victories.

~ Abraham Lincoln (1809-1865), 16th President of the United States

To maintain quality while performing on a schedule, ensembles and innovation teams need vigilance. Vigilance takes the form of repetition, pacing, and responsiveness. On each of these matters, vigilance depends on many personal decisions; a competitive spirit brings out each team member's best decisions and contributions.

In the arts, performances are measured in minutes. All the other time—practicing and rehearsals—is measured in hours, days, and even weeks. The rehearsal period typically has several phases, and the ensemble must vigilantly meet every interim milestone to be performance-ready for their audience.

Innovation teams have a similar time distortion. A project can easily last ten times longer than a formal audience launch. With their own phases, an innovation team must vigilantly meet every interim milestone to be customer-ready and employee ready with new processes.

Repetition

All the arts benefit from their respective forms of drills and repetition. Parents set routines for their children and themselves. Musicians play scales and perform warmups. Dancers forget what they learn if they don't use it on the dance floor. Repetition results in performing with less effort, higher quality, and muscle memory.

Healthy repetition in innovation includes executing Five Verbs for a stable asset portfolio across projects and producing weekly status reports. A lack of repetition feels like fire drills and chaos.

Pacing

Development and improvement of artistic material are gradual, not sudden or automatic. Rehearsal must be focused and intentional. Martial artists learn kicks, pulls, punches, and throws in a conscious order. Parents give their children age-appropriate experiences. Dancers learn mechanics before style. They pace steps to match the tempo of the music.

Any ensemble member might learn slowly, struggle with a small portion of the material, and need extra rehearsal time to keep up with the schedule. Other members must give that person grace while they work extra hard to fulfill their obligations. Feedback is vital, frustration is common, and awkwardness is inevitable. But impatience for instant excellence is counterproductive. Rushing things often leads to poor quality, stress, or injury.

In innovation, impatience is also a recipe for disaster. Unfortunately, it's a cliché for an inexperienced team to overpromise on scope or a schedule. An impatient team skips documentation, skips consulting stakeholders, and skips approvals.

It's also counterproductive to procrastinate. Procrastination undermines memorization, muscle memory, and the enjoyment of performance. Procrastination with solo projects is difficult enough, and procrastination with team projects disrupts others' schedules (and as a result is next to impossible to overcome). A good leader in the arts and innovation ensures their team has a comprehensive, integrated, and well-paced schedule. When the team has faith and confidence in the schedule, everyone works hard to meet interim milestones and minimize negative surprises.

Procrastination often leads to poor quality, disengagement, or resentment. The straightforward way to avoid impatience and procrastination is pacing. Healthy projects, like healthy performances in the empathetic arts, have pacing and rhythm.

Responsiveness

Another form of vigilance is the responsiveness team members have moment-to-moment. The arts have high interdependence among the performers; vigilant performers are ready to contribute when the flow of work comes to them. Kids need their parents multiple times per day. An improviser waits for the dialogue to come to them. Dance partners continuously send and receive signals from each other. Performers are best prepared for surprises when they have an alternate script ready to execute, have seen the surprise before, and feel prepared to improvise.

A diligent innovation team member knows when the team needs their contribution and is responsive to it. A schedule with Five Verbs provides weeks of visibility in advance, but surprises can happen. A healthy innovation team, therefore, is vigilant in the same ways. For example, they execute in a script-like manner, know what surprises are possible, and are always ready to pause, pivot, and improvise.

Performers accept that the arts require vigilance; the risk of surprise always exists and justifies caution and contingency. Orchestra members keep their instruments clean and lubricated. Actors have understudies in case someone cannot go on stage. Martial artists cannot let their guard down. Experiencing even one negative surprise, like an actor missing a single cue, reminds us how fragile teamwork can be.

As an opening night performance approaches, you must rehearse as you will perform, minimizing differences between your rehearsal and the actual performance. This kind of vigilance reduces the inevitable stress and stage fright, raising the confidence and enjoyment of all stakeholders.

For innovation, the project is the practice. Like the arts, healthy practice and rehearsal for your innovation team require vigilance in the form of repetition, pacing, and responsiveness. Innovation teams exercise vigilance by having the right team members contribute to the proper documentation. For example, weekly status reports and an issue/risk log are vital to show vigilance. These two assets exercise transparency, maintain a healthy tempo, and minimize negative surprises.

Vigilance is never easy, and you only control your own. Vigilance requires personal ambition. It is a competitive advantage for a team. Vigilance is the cost of staying engaged, safe, and relevant.

Rejection

Go that way. Really fast. If something gets in your way, turn.
~ Charles De Mar, Better Off Dead (film, 1985)

Embracing a culture of rejection is counterintuitive, but teams and individuals skilled in rejection have advantages. Rejection is a valuable, admittedly unstylish, skill. Embracing rejection is evidence you recognize your self-worth, have humility, and are out there trying. Avoiding rejection is evidence you avoid your best opportunities—effectively choosing rejection for yourself before anyone else might choose it for you. Rejection is a part of many processes, especially when you are exploring new things. Rejection shows you are confident in your competitive advantages.

Rejections permeate the arts, making them models for rejection skills. Parents mentor their children through disappointments such as not making a sports team, dealing with friendships that end, and with relationships that never start. Musicians and actors experience disappointments via many auditions.

For innovation professionals, rejections take the form of interviews, promotions, and downsizing decisions that don't go your way. Innovation professionals are happier when they recover from rejection as skillfully as experienced auditioners do.

It's natural to look inward and take rejection personally, but you rarely know others' circumstances that lead to rejection. To succeed, you don't need to hear "yes" multiple times. Just as movie stars shoot one movie at a time, you are limited to one "yes" at a time. The point is to spend time and energy on what you can control—in other words, choosing and preparing for opportunities despite fear of rejection.

The skill of rejection has two sides: giving and receiving rejection.

Giving Rejection

An expert rejects another person with discipline and empathy, aiming to minimize the risk of vindictiveness. Parents are the epitome of giving rejection since they say "no" to their children in countless ways. Good parents calibrate the intensity of their rejection according to the potential damage of what the child wants to say or do.

Innovation professionals do everyone a favor when they calibrate their rejection according to the potential damage of what an employee wants to say or do. An expert at giving rejection draws intelligent boundaries. Rejection skills accompany positive self-esteem, fearlessness, and focus.

Receiving Rejection

Similarly, an expert receives rejection with poise. A rejection expert rarely feels surprised and has a 'thick skin' about rejection. An expert captures any available lessons and moves on.

The arts have many rejection experts because of their culture of rejection. Martial arts are the epitome of receiving rejection since every motion is an attack. Ignoring discipline or empathy, a novice receives rejection with an emotional or demonstrative meltdown. A rejection novice feels surprised more often and might have a 'thick skull' about negativity toward them.

Innovation professionals do everyone a favor when rejection is not surprising. When innovation professionals emulate martial artists, they grow a thick skin and become experts at receiving rejections.

The skill of rejection applies to individuals and teams alike. An unskilled individual avoids others, often working in isolation and fear. They skip auditions and don't ask anyone to dance. A skilled individual instinctively gravitates toward people but respectfully backs off when they sense someone else is uncomfortable; they audition assertively and work hard for everyone to retain dignity in the face of rejection.

An unskilled team has poor boundaries, poor prioritization, and is trapped in mediocrity. A skilled team picks its battles, keeps a healthy workload, and proceeds cautiously—like good parents who keep their children away from unhealthy activities and influences.

Some rejection is sudden and occurs in a single event. Another form of rejection—being diminished—is slow and occurs across several weeks and events. In the arts, one performer might diminish another when they see the other as a threat to their status or personal success. The diminisher seeks to make the other person so miserable and unwelcome that they want to leave the ensemble.

In innovation, diminishers lack the discipline and empathy to be transparent about their disapproval. They undermine a team member's morale and performance. Diminishers hurt the value of team members and the team itself. A silver lining about a diminisher is that their rejection is not a surprise.

Some rejections are reflections of the person giving the rejection. Some rejections are reflections of the receiver. Pay attention to any repetition of rejection. Any person experiencing a pattern of rejection should reflect on the patterns and rethink their approach. A pattern of giving rejections suggests the person is being overly picky. A pattern of receiving rejections indicates the person is not learning why they're a poor fit and/or what they can do to change to avoid future rejection.

Rejection is often a blessing in disguise. When one party lacks the awareness or courage to end an arrangement, the other's rejection can be a merciful escape from mediocrity. Rejection is a lubricant that helps you move on to better things. It's courageous to end an arrangement before its value dramatically erodes. For example, in parenting—more so in extended families—subtle distancing can be better than complete estrangement.

Rejection novices often don't realize this, and they get hurt twice. The first hurt is the end of the relationship. The second hurt is the time spent dwelling on the change. Rejection experts give themselves the grace to spend as much time as they need for grieving, but they are resilient and move on as quickly as they can. A skilled musician can learn of an unsuccessful audition one day and nail a successful audition the next.

Smart rejections free you from the places where you don't want to spend your time. Smart strategy is knowing how not to spend your time. Rejections encourage exploration, improvement, and innovation. The clarity in rejection is a license to move on.

When a decision maker says something is a bad fit or not a priority, accept their authority (and accountability), and pursue alternate projects and assignments. Attractive alternatives[11] whether ideas, employers, or customers, are valuable. Great alternatives make it easy to experience rejection.

Rejection skills are not glamorous or high-profile, which makes them even more of a distinctive competitive advantage. Procrastinating rejection traps you in mediocrity. Flowing with rejection propels you into scenarios with higher appreciation and potential. Short-term rejections improve the long arc of business and life—in the arts and innovation.

Ignoring rejection skills leads to a fragile team, fragile individuals, and a victim mentality. Rejection skills unleash learning, resilience, and courage to pursue your full potential. The arts push you to show up, audition, and be courageous when facing the prospect of rejection. Innovation professionals benefit from the same courage artists display.

Failure

Success is not final, failure is not fatal: it is the courage to continue that counts.

~ Winston Churchill (1874-1965) Prime Minister of the
United Kingdom 1940-1945; 1951-1955

Rejection's sibling is named Failure. If you are good at what you do, you will fail because it means you are out there taking risks. When creating something new, there will be small failures along the way. Organizations must plan for those failures in ways that will not put the entire enterprise at risk. Businesses need elasticity to recover. In the face of failure, the ability to recover is a competitive advantage for individuals and teams.

Conventional wisdom tells us that failure is terrible. Hard-nosed managers warn, "Failure is unacceptable!" These statements are untrue and unhelpful. What's terrible is the *fear* of failure since it can paralyze you. Risk-averse

---

11 Negotiation best practices promote the value of a strong BATNA: Best Alternative to a Negotiated Agreement.

organizations become stagnant and obsolete. When an organization loses its ability to take a risk, it is the start of artistic death.

The arts contain many forms of failure. A martial artist can fail to defend themselves or earn their next belt. A dance couple might try choreography for a few weeks before concluding it is too difficult to pull off reliably in a performance. A musical ensemble might rehearse a piece of music before concluding its harmonies are unpleasant. As Churchill says, none of these 'failures' are fatal. Artistic death is *fearing* these experiences and experiments.

To avoid this fear, the arts contain ways to anticipate and proactively manage failure. A dancer might test some choreography before committing to it. Likewise, an innovation team must proactively manage for some instances of failure. This isn't the same as surrendering to failure. It means you take steps to minimize the quantity and severity of failures. The Elegance methodology aims to accomplish precisely this by minimizing project failure rates.

Failure for an innovation team is ambiguous. It takes many forms because, even for a single project, different stakeholders might define or describe failure differently. Some failures are objective and measured by revenue, cost, or schedule. Some failures are subjective, expressed in terms of customer satisfaction, employee morale, or turnover.

Some failures are those that miss realistic expectations, such as a 10% customer increase. Some failures result from *un*realistic expectations, such as a 500% customer increase. And some failures miss on perception; for example, listening to a vocal minority of disgruntled stakeholders instead of a less vocal majority of pleased customers.

The Elegance methodology proactively manages for these forms of failure through the asset portfolio and how its transparency governs unambiguous small successes and failures along the way. The discipline and empathy of the asset portfolio shape a team's self-esteem, humility, and fearlessness regarding failure. Elegance accepts small failures (sampling errors) and elegantly moves on. Elegance avoids repeated systemic failures and purposeful systematic ones.

The arts help innovators see failure in a positive light. The early arrival of a minor failure means your ensemble stopped investing in something unrealistic or of low value. Small failures are teachers, often teaching us to do something else. Transparency of options eases recovery and helps to move on from a failure. A small failure—in the form of trying, learning, and moving on—is hardly a failure. True failure is not doing any of these things. Small failures are symptoms of ambitious experiences.

The arts contain many examples of small failures that steer an ensemble to long-term vibrancy. Parents might fail trying gymnastics or piano lessons for their children, and those failures can lead to swimming and guitar lessons. An experienced saxophone player might abort an experiment to play a different instrument. An innovator might have their first idea flop but continue as part of a team that conceives and launches a customer breakthrough.

Arguably contradictory, innovation teams must accept some failures *and* work hard to minimize them. Detecting and responding to small failures prevents big ones from happening. Sometimes an accumulation of failures is required before you succeed.

In the short term, failure stings. In the long term, courage in the face of failure is a competitive advantage.

Resilience

The road to success is paved with mistakes well-handled.

~ Daniel Coyle (b. 1965), author, The Talent Code and The Culture Code

Bouncing back from repeated rejection, failure, and negative surprises requires resilience. Artists have resilience in abundance. To similarly combat the threat of VUCA, innovation teams need the competitive advantage of high resilience.

Parents don't give up on their kids. Martial artists don't surrender at the first, second, or third attack. Dancers keep trying moves they don't get at first. Without resilience, these experiences abruptly stop.

Resilient innovation professionals have grit. They persevere through inevitable surprises. Every project, team, and culture has its dysfunction. Whether you stick with a rough situation or make a fresh start, it's often only a matter of time before you face another setback.

Some parents seek advice from their parents. Improv groups typically have coaches. Resilience is easier when you're willing to ask for help. People who find success always receive help along the way. There's a good chance that anyone you ask for help has had a similar experience, and they're now in a position to help others. Consider that when you are not in immediate need of help yourself: it's easy to give help or advise others to get it.

Many innovation professionals have low resilience due to pride, a victim mentality, or an unwillingness to ask for help. You can improve your resilience if you prepare for negative experiences. Preparation includes remembering that many people have already had the same experiences you will have. Many experiences are not within your control—reducing self-blame helps resilience. Avoiding a victim mentality prevents the paralysis that stops you from moving on.

Some innovation failures result in job loss for many people. The root cause of these failures is not junior employees. The root cause of every innovation failure is a few senior employees—possibly even a single executive. Keeping this firmly in mind should help junior employees build resilience with every innovation failure they experience.

Consider what psychological insurance you need to minimize damage from a setback. Psychological insurance can include people you trust, a robust professional network, and ample career options.

Arts organizations need resilience, too. A music director keeps contact information for alternate rehearsal spaces and musicians. A dance team rotates partners so that an absence doesn't cripple the team. Resilience helps each organization stay competitive among its peers.

Resilience for an innovation team requires simplicity and transparency of its methodology. A poor methodology weakens a team's resilience to negative surprises. The Elegance methodology emphasizes documentation to foster transparency and manages through Five Verbs to enforce simplicity.

A bank account is an excellent analogy for resilience. Resilience is the ability to make a surprise withdrawal without bankrupting the account. Preparations for a surprise withdrawal are deposits. Deposits are a form of upfront cost. Withdrawals are a form of marginal cost.

Resilience in a relationship requires deposits of trust so that the relationship can survive the withdrawal of a negative surprise. Therefore, resilience to losing a team member requires investing in more people than the bare minimum so that your team can survive a surprise departure. Resilience in the innovation workstream requires innovation alternatives so your team can survive the cancelation of a project. If you pay these upfront costs, your marginal cost is low.

Resilience is the road to success paved with mistakes well-handled.

The Artist-Audience Intersection

Both artist and audience make up a work of art.

~ Helen Martineau, Australian author, storyteller, and metaphysician

The previous two sections presented culture traits in the arts that shape and distinguish collaboration and competition among the artists. This section explains the culture traits that shape the audience experience and the intersection of the artist and the audience.

In the arts, the conventional arrangement is that the performer provides time and talent, and a passive audience pays for the show. In business, the traditional arrangement is the innovator provides time and talent, and the customer provides payment. But these relationships vary among the arts, from business to business, and they change over the years.

Traditionally, as performers prepare a concert or a play, they shelter the audience from the complexities of preparation. Innovators call this a turnkey, that is, something ready to use out of the box. This arrangement creates distance between the innovator and their customer. It is low maintenance and low intimacy. Customers don't meet their innovators to experience the final product or service, and audience members don't meet the artists.

Some artists welcome feedback and ideas from audiences. Improv teams, for example, ask for ideas during the performance. These artists ask audiences to be more active and less passive. Increasingly businesses do the same thing. They don't just ask for payment; they ask for the audience's time and talent. Some businesses even 'pay' their customers for ideas and survey responses with credits for additional products or services. These companies consciously choose high customer intimacy over low customer intimacy.

The highest customer intimacy is when artists completely blur the lines with their audience. Parents play with their kids instead of disciplining them. Audience members are invited on stage. Compliments flow in the opposite direction when artists exclaim, "You've been a great audience!" The audience might shape the performance, provide some performers, and seamlessly collaborate with the artists.

In scenarios of low customer intimacy, feedback and judgment flow one way, i.e., from the audience to the artists. In the scenario of high customer intimacy, judgment can flow two ways, i.e., how artists experience the audience. The adage 'the customer is always right' was not everyone's reality in decades past, and it's not everyone's reality today.

In the arts and innovation, organizations spend a lot of time on performance feedback. This is critical to improving performance while managing expectations. Some organizations emphasize numbers and track metrics that matter in a scorecard. Some organizations emphasize commentary, and track moments that matter in people's minds. Both forms of audience feedback influence innovation for artists and businesses. Likewise, an artist's empathy for their audience and a business's empathy for its customers contribute to innovation.

Ideally, the artist-audience relationship lasts as long as it is uniquely valuable. Whether the relationship lasts a couple of performances or a couple of decades, it comes to an end. Audience members have fond memories, a favorite melody, and a favorite scene to keep. Artists take forward their portfolio of work with each performance adding to their legacy in their craft.

Innovation professionals have a similar scenario. Ideally, they contribute to a customer experience as long as it is uniquely valuable. Customers have an improved portfolio of processes, products, and services to keep. Innovators have improved skills, resumes, and reputations to take forward.

Competitive forces have always shaped the business world. But society is demanding more empathy from it. That empathy shows up as collaborative innovation among employers, employees, customers, and extended stakeholder groups. Increased empathy shifts emphasis from competitive work toward collaborative work. Collaborative outcomes are higher quality, more fun, and more profitable. The world of work is becoming more intense, making companies hypercompetitive *and* hyper-collaborative.

As hyper-collaboration becomes more valuable than hyper-competition, those forces might lead the world of work to resemble one big team. Innovators and customers are set up to thrive when they hyper-collaborate—imitating the artist-audience intersection.

GETMO

> *The pursuit of excellence will motivate you. The pursuit of perfection will eventually limit you.*
>
> ~ Craig Groeschel (b. 1967), Oklahoma-based pastor and author

In the arts, the audience has certain expectations of the performers. Performers have expectations of themselves, too—often higher expectations than the audience. Perfection is a common problem for artists. No one has unlimited time and, whether an expectation of perfection comes from performers or an audience, perfection is usually a counterproductive goal.

Innovation teams face the same risk of perfectionism. Mature innovation professionals know and manage customers' expectations for speed, quality, and cost. Because perfectionism hurts all three. Good innovation leaders optimize across numerous expectations and can recognize when work is GETMO: Good Enough To Move On.

Art ensembles and innovation teams benefit from attention to detail. Precision conveys high quality and commands high prices. But teams can try for too much of a good thing. Perfection is unnecessary and unrealistic. Quoting the 18th century French writer Voltaire, "Perfection is the enemy of the good." Instead of making one flawless piece of work, think of perfection as expertly shifting your attention from one activity to the next because your time is more valuable there.

GETMO exists across the arts. A music ensemble is an excellent example of a team repeatedly declaring GETMO. A symphony's performance date is typically fixed, so during rehearsals, the ensemble must prioritize what to improve and accept some things that feel less than flawless. And if a mistake occurs in a performance, MO is the appropriate acronym because the music ruthlessly moves on.

Healthy innovation teams also execute GETMO to avoid becoming paralyzed in one activity or asset. GETMO is not strictly a reflection of the current work's quality; it reflects that the value of additional time and effort on the current work is low, and that the value of the next task will be higher.

GETMO applies to Five Verbs. A team member should spend a comfortable amount of time drafting an asset. As soon as the asset is GETMO, the drafter hands off the asset to other team members. Those team members should also spend a comfortable amount of time reviewing and revising it. As soon as the asset is GETMO, they should hand it off to the team members assigned to approve it.

Every action in your innovation factory has a tipping point where the work is GETMO. Don't think of perfection as the condition of the asset. Think of perfection as transitioning at a healthy, sustainable pace. Once your team nears exhausting their collaboration on one asset, acknowledge the so-called 'point of diminishing return.' Approve, distribute, and move on to the next asset.

GETMO is a reminder that, even without world-class talent, you can still have fun and be productive. Often, just doing the fundamentals well is good enough. Some teamwork is truly just average, and average is not a crime. Handling mediocrity with dignity and grace is still more profitable than being absent or excluded.

When artists and innovators have high standards for themselves, they commonly notice imperfections that audiences and customers never will. It requires discipline and maturity to accept flaws to minimize jeopardizing other aspects of the work, especially a schedule.

A healthy innovation team accepts imperfection gracefully and acknowledges *continual* imperfection among customers and within the team. The team continually improves the experiences its stakeholders value most. It fixes systemic and systematic errors for its customers and within the team.

These culture traits help a team's value far more than their sampling errors hurt. Spotlighting sampling errors is petty when a healthy team excels at prioritizing and fixing matters of enormous financial consequence. GETMO helps teams avoid conventional perfectionism and instead treats perfection as always concentrating on the transitions in advancing the project's most valuable assets.

A mature audience has realistic expectations, and they'd rather see a GETMO performance (in the eyes of the performers) than wait for unrealistic perfection. Innovation teams can operate with the same expectation. At some point—earlier than a perfectionistic team realizes—all innovations are GETMO.

Performance

> *It's not how you start. What matters is how you finish.*
>
> ~ Jim George (b. 1943), American seminarian

In music and theater, rehearsals are the journey, and performances are the destination. Opening night is festive, and the artists' confidence is high. The hard work is over, and artists and audiences now reap the benefits. When artists deliver a good performance, the audience shows appreciation during and after the performance. The performance is the climax of the artist-audience intersection.

In innovation, the project is the journey, and a go-live event is the destination. When the project team has executed a sound methodology, confidence is high, and the mood is festive. The hard work is over, and employees and customers reap the benefits. When the improved experience is in place, customers express their appreciation. The go-live event is the climax of the innovator-customer intersection.

Of course, performances are rarely perfect. Musicians and actors are human and make small mistakes. Performers might beat themselves up or not enjoy the performances as much as expected. But blemishes aside, a high percentage of artists and audiences are eager for more. Enthusiasm for more, however unscientific, is compelling evidence of high success rates in the arts.

Innovation go-live events are rarely perfect. Team members might be nervous, pessimistic, or burned out from working long hours near the end of the project. The percentage of employees and customers who are eager for more seems lower than the similar metric in the arts. The evidence is the overwhelming amount of dialogue in the business community about shortcomings in leadership, management, and methodology.

Improving innovation's success rate by focusing on leadership and management is an ambiguous, long-term investment. Improving the success rate by focusing on methodology is an unambiguous, upfront investment.

Innovation Elegance leverages six empathetic arts for two reasons. First, each art already has culture traits and a methodology for high success rates. Second, when an innovation methodology leverages all the arts together, success rates will leap-frog their current ones. Innovations teams will approach their go-live event—their opening night—with extraordinary enthusiasm and confidence.

The climax for artists is the performance. The climax for innovators is the go-live event. In both worlds, tools for discipline and empathy yield an elegant methodology, boosting performers' enthusiasm and confidence in their prospects for success.

Celebration

We were together, I forget the rest.

~ Walt Whitman (1819-1892), American poet and journalist

Artists know how to party. Most innovators do, too. Innovation teams must celebrate their big wins for employees to enjoy the journey, feel appreciated, and have a sense of accomplishment.

Parents celebrate their kids' first steps, birthdays, and graduations. As a theater audience applauds a play at a curtain call, the cast redirects attention to stage-crew members. At a concert's ovation, the music director typically diverts attention to soloists and special performers.

Similarly, an innovation leader who gets enough financial and professional recognition for themselves celebrates the junior employees by directing credit to them. For large projects, the sponsor or CEO commonly broadcasts a thank-you email, mentions the team at an all-company meeting, or makes a speech at the project party. Many companies celebrate new products with advertising, a press release, and a party for their top customers. Every company should celebrate in ways that fit its culture and finances.

In the arts, as performance runs near completion, the tone shifts from celebratory to sentimental. It resembles academic graduation—some friends depart for other opportunities and other performers stay with the ensemble, learn what's next, and begin the performance lifecycle again. Whether actors stay or go, the arts benefit from the enthusiasm for more. There is always more music to make, dances to learn, and stories to bring to life.

Innovation teams have the same excellent problem: there is always more to innovate. The employees who continue working in the organization determine what to learn and innovate next, and the innovation lifecycle begins again. Ideally, innovation professionals go where their growth opportunities and career security are best.

Because of the arts' enthusiasm for more, artists are often sentimental about their past teams and colleagues. This sentimentality is valuable to future ensembles because it makes future collaborations easier. Artists leave each other with positive reputations—in stock market parlance, they 'sell with their stock high.' Artists keep in touch and are resources to each other.

With the Elegance methodology, innovation professionals can shape cultures of similar enthusiasm, positive reputations, and project success rates. Teammates can move on 'with their stock high.' A software-centric innovation methodology jeopardizes projects, jobs, and relationships, while a people-centric methodology nourishes them.

In the arts, audiences are generous with their appreciation for the artists. As innovation professionals execute a methodology inspired by the arts, customers and extended stakeholders will likewise be appreciative and celebratory.

Moments That Matter

Life isn't a matter of milestones, but of moments.

~ Rose Kennedy (1890-1995), matriarch of the Kennedy family

In the arts, metrics matter: audience size, sold-out shows, and quantity of awards. To its credit, quantities are objective, and many decisions in the arts need objectivity. But the arts draw audiences because performances create memorable moments for them.

Metrics also matter for innovators. But innovators distinguish themselves when they create moments that matter for their customers. Innovation professionals distinguish themselves when they create moments that matter to their peers.

All the empathetic arts generate moments that matter. Dancers have moves and poses that make audiences erupt in approval. Symphonies have majestic walls of sound that audiences replay for decades. Movies and theater inspire lines that show up in everyday conversation for the rest of people's lives. These moments have an outsized effect on shaping the audience experience.

The arts can—and do—inspire moments for businesses' customers to shape the customer experience. Customer experience is valuable enough for businesses to justify a profession for it. One formal definition of customer experience is "the cumulative impact of multiple touchpoints over the course of a customer's interaction."

Instead of strictly selling products or services, many companies create experiences to delight customers, increase customer volume, and justify higher prices. Businesses stand to thrive looking to the arts to inspire moments that matter and create exceptional customer experiences. Beyond great taste, foodies love the restaurant experience. Families want their kids to meet mascots in costumes. Car dealers greet their customers with a glass of champagne.

In the arts, a performer's experience differs from an audience's experience. If the culture shapes a negative experience for performers, they look for alternative theater companies and dance studios. A culture that creates a positive artist experience attracts top talent, possibly at a modest cost because of the welcoming environment. It's the same in innovation. A negative employee experience motivates professionals to look for alternative employers. A culture that creates a positive employee experience attracts top talent, often at a modest cost.

As a company continually improves the customer experience, it might focus on the potential upsides for its customers, i.e., 'points of delight.' The upside might add a new positive experience or re-engineer an existing, mediocre experience to make it unforgettable.

Alternatively, a company might focus on re-engineering negative experiences to make them less painful. Moments that matter typically refer to maximizing delight, but there is value in minimizing points of pain.

Empathetic leaders make conscious innovation choices along a spectrum of pleasure and pain—delight and despair.

Metrics matter. But businesses have unlimited potential when they model the arts and create moments that matter for both customers and employees. These moments are often inexpensive to engineer, but with payoffs that are high and enduring.

Because of personnel changes, companies can only partially rely on managers and leaders to create these moments. The most empathetic companies automate creating moments that matter. Systematizing this relies on your innovation methodology. Companies distinguish themselves when they have a culture that continually re-engineers for moments that matter.

Legacy

The idea is not to live forever, but to create something that will.

~ Andy Warhol (1928-1987), American visual artist
and leader of the Pop Art movement

Companies change. People move on. What they leave behind is their legacy. Innovation professionals aren't typically coached—as artists can be—to work toward their legacy, but such coaching inspires, expands stakeholder impact, and is more fun than not working toward a legacy.

Reputation is what others say about you when you are not around. Legacy is what others say about you when you've permanently left the scene. Non-empathetic people mostly care about themselves, and that is their legacy. Empathetic people care about others, and thus caring is their legacy.

The arts are full of artists with legacies. Parents leave a legacy for their children. Musicians, dancers, and actors populate celebrity culture. They leave behind catchy melodies, dance fads, and movie lines. Artists don't live forever, but their work does. Artists build things that are bigger than themselves.

Innovators also leave legacies, whether large or small. Not everyone can be Steve Jobs or Leonardo Di Vinci, but every innovation professional impacts their team. And every team impacts its customers. Innovators leave behind experiences for others to pick up.

At the humble level of a project, a project team leaves a legacy in the plain form of documentation—the asset portfolio. After the project team moves on, the asset portfolio is what subsequent teams talk about and must work with. A team with low empathy doesn't leave documentation. A highly empathetic team leaves documentation that makes the subsequent team's work easier than without it. Documentation is a gift to a future team. That future team is the audience for your current team.

That asset portfolio warms the heart, and it also warms a company's finances. All the upfront work benefits the next team because the transparency of the business enables innovation to proceed faster and cheaper. In comparison, teams with low empathy and documentation raise cost and slow down innovation.

Artists profit from their portfolio, and their portfolio shapes their legacy. The portfolio makes it cheap and easy for future generations to enjoy their portfolio repeatedly. Innovators can learn from this. Innovators profit from a project team's portfolio and legacy. The portfolio makes it cheap and easy to innovate things bigger than themselves.

Distinct Experiences within Each Empathetic Art

The artist sees what others only catch a glimpse of.
~ Leonardo Da Vinci (1452-1519),
Italian painter and scientist

The preceding sections outlined the culture traits that the arts share, how innovation teams already emulate these traits, and how innovation teams benefit by deliberately emulating the arts. These traits are valuable because they enforce collaboration alongside competition, provide a people-centric methodology, and set innovation teams up for success.

The following sections reflect on the distinct culture traits within each of the six empathetic arts and how innovation teams can benefit from their language, habits, and culture. Similarly, to the previous sections, your team might already emulate much of what you read. But you stand to benefit from whatever you don't yet emulate. If exploring six arts feels overwhelming or counterproductive, select one or two that resonate with you.

These six sections are not meant to be exhaustive; rather, each section shares how you can connect the dots between these six arts and your innovation team. Whatever connections resonate with you, apply the language, habits, empathy, and most importantly, grace, to your team.

Parenting

A good parent's life is not their own.

~ Unknown

Over the past few decades, employer and employee relationships have changed dramatically. The distant past of company culture included tenure, loyalty, and job security. Recent company culture includes cut-throat competition, mass layoffs, and fire-by-tweet. Too often, companies care about employees for the short term rather than the long term.

One stereotype of innovation culture is that it is cruel and uncaring. Employees own their development, networks, and career. Succession planning typically exists for senior-ranking employees, but the equivalent for junior employees—growth, self-sufficiency, and a talent pipeline—is not standard. And because junior employees include numerous people who could outperform incumbent executives, insecure senior employees might have an incentive to keep the next generation from blossoming too quickly. To judiciously govern your team, you need skills and tools to protect your employees, develop them, and set them free.

Parenting is a distinctive art because of the complexity of its interactions. Healthy interaction involves pacing a child's autonomy and independence

157

over decades. It involves extreme discipline and extreme forgiveness. Good parenting accepts and recovers from awkward, messy, and horrifying events.

When asked, "What is your leadership style?" good parents have two answers: "Bang the table," and, "Set the table." Good parents are mindful of occasions when a child is vulnerable to a bad decision and negative consequences so should have no authority over what they have in mind. Good parents are also empathetic to where a child's autonomy and accountability are more important than the negative consequences of their decisions. Good parents allow their children to make some of their own mistakes.

Parents pick their battles along the journey with their children. They manage age-appropriate matters such as allowing ten-year-old children to play tennis or baseball for twelve hours straight, or allowing teenagers to wallow in their first broken heart.

A parenting metaphor is valuable, although it can be abused. It can be an excuse for favoritism, control, and a dysfunctional power imbalance. Parenting extremes led to terms such as 'absentee father,' 'tiger mom,' and 'helicopter parent.' But for innovation professionals who can strictly look at the metaphor for its intended benefit, parenting provides a unique model for empathy, discipline, and developing others to be self-sufficient team members.

Good innovation leaders emulate the best aspects of parenting. They care without coddling or forever controlling. They coach without crushing budding independence. They balance a team's cohesiveness with each employee's personal goals.

Although an imperfect metaphor, references to 'family' are common enough in the innovation culture to defend how the empathy of parenting is a business strength and advantage. High turnover and short-term thinking are an expensive disadvantage. Parenting is a model to invest in for employees' long-term viability and health.

Bang The Table

It is easier to build strong children than to repair broken men.

~ Frederick Douglass (1818-1895), American social reformer and abolitionist

Parents exert high levels of control over a child's first few years of life. Parents control what a child sees, touches, hears, and eats. A vigilant parent keeps a child safe and their autonomy low.

New members of an innovation team can have a similar experience when their boss exerts strong control. Children and inexperienced team members have little say in what they see and do.

Between the ages of two and six, a child has learned enough to push boundaries. Parents shift their attention from safety to education and setting expectations. Rules resemble a bang-the-table leadership style. Parents constantly choose along the spectrum of hands-on versus hands-off leadership.

An innovation worker has a similar experience. A wealth of initial learning, curiosity, and ambition leads a typical team member to push boundaries after a while. After an early honeymoon phase, it's common for a worker to have friction with colleagues and their boss. A worker might feel micromanaged or abandoned. Ideally, both are rare.

Inevitably, children reach an age where they push their independence too far, occasionally or repeatedly. To bang the table and enforce their wishes, parents say, "Do as I say," "Not under my roof," and "This family is not a democracy." Parents might revoke privileges as a form of bang the table. Parents feel they know the difference between right, wrong, and what's best for the family. Healthy parents exercise their accountability and authority.

Inevitably, employees reach a tenure where they, occasionally or repeatedly, push their independence too far. Innovation leaders are empowered to govern what's best for their teams and customers. They are empowered to 'lay down the law' and convey 'it's my way or the highway.' Healthy innovation leaders exercise their accountability and authority.

A less dramatic version of bang-the-table leadership is when a parent or innovation leader serves as a tiebreaker. Children might prefer different pizza toppings, and an adult might need to serve as a tiebreaker. Innovation team members might give diverging recommendations, and a leader might need to serve as the ultimate decisionmaker.

Bang-the-table leadership resembles the command-and-control management style. Over the past few decades, it has become less effective and less common, but occasions exist in innovation that resemble a five-year-old child running into a street full of oncoming cars. When a team member is about to make a big mistake—perhaps a poor ethical decision—bang-the-table leadership is the right way to lead.

Set The Table

Man plans and God laughs.

~ Yiddish proverb

Parents gradually reduce their oversight as children grow, and formal schooling increases its influence. Parents move from bang-the-table leadership to set-the-table leadership. Children have options, and parents are there to facilitate them. Options arrive in the form of activities such as sports, art, and music. Children specialize in activities they enjoy and shine. Teachers and coaches take on significant leadership roles in their lives.

Innovation managers calibrate autonomy for their employees. A good manager prescribes the work's 'why' before deciding what to oversee or delegate to the junior employee.

- Low autonomy oversees what and how of the employee's work.

- Medium autonomy oversees what.

- High autonomy oversees neither.

Facing more choices, a teenager's honesty and integrity matter. Ideally, teenagers choose good study habits and good friends. Of course, some teenagers choose poorly, rebel, and find themselves in trouble with their parents, school, or the law.

Similarly, more experienced business professionals have high latitude in many decisions. The boundaries of behavior matter, and dishonesty can have extreme consequences. Some professionals take so much rope that they ethically hang themselves.

As teenagers reach driving age and legal adulthood, their autonomy continues to grow. They see, feel, and own the benefits and consequences of their decisions. Good decisions land awards, scholarships, and job offers. Bad decisions squander opportunities. However sheltered a child might be, the teenage years expose them to more positive and negative role models, and parents can no longer childproof their child's life and decisions.

An innovation employee's exposure to options also increases. Once they feel they've mastered one job, they look for something new. They can accept new assignments, teams, and employers. They are less sheltered and are more confident in what they want. Innovation employees typically have great options for their career direction, whether in business analysis, change management, or technology and data. The universal need for innovation should minimize feeling trapped in a job that isn't fulfilling in ways that workers want to be challenged.

Messiness

Pardon the mess. My children are making memories.

~ Unknown

Even with the best combination of habits and choices, parenting is messy. When the military conceived the acronym VUCA, they could have easily been referring to family life. The innovation world has since adopted the term VUCA as an explanation for messy projects and an excuse for inelegant innovation. But just like the most successful families have ways to counter VUCA in their lives, so do the best innovation teams.

Neither parents nor innovation professionals can completely avoid messiness. Many things outside their control impact themselves and the people under their care. What parents can control is their preparation for messiness and their reaction to it. And it's possible to over-prepare, under-prepare, overreact, and under-react. A leader's tolerance level for messiness influence decisions for preparation and reactions.

The nature of preparation and reactions evolves over time. With a young child, messiness takes the form of crying, flying food, and diapers. A good parent is naturally prepared for this, and their reaction is forgiving—low discipline and high empathy. With a new employee, messiness involves questions, presumptions, and mistakes. A good leader is prepared for this, and they react with grace. Young children and new employees need some figurative and literal hand-holding.

As a child ages, the forms of messiness change into hurt feelings, an unwillingness to share toys, and a messy bedroom. A good parent changes their preparation and reaction to suit the child. A good parent proactively sets and manages expectations with the child and is less forgiving—medium discipline and medium empathy.

For an experienced innovation team member, messiness takes the form of disorganization, poor collaboration, and personality conflicts. A good leader tailors their reaction, firmly sets expectations, and is less forgiving of messiness. But to a significant degree, the leader still meets their followers where they are.

A third phase of messiness is when the child creates messes for other people. Aggressive behavior, such as bullying and violating boundaries, physically and psychologically damages others. The parent of a bully prepared their child poorly. The bully, and possibly the parent, deserve high discipline and low empathy. Victims and their parents must prepare and react to the messiness imposed upon them.

Unless governed well, innovation team members can create messiness for their colleagues. This also takes the form of aggressiveness and toxic competition. Leaders and managers of toxic competition neglect their employees. Predatory employees, and possibly their boss as well, deserve high discipline and low empathy. Victims and their managers must prepare and react to the messiness imposed upon them.

Another type of messiness is when the parent is the problem. The messiness is extreme neglect, toxic discipline, and insincere empathy. The parent exhibits little or no healthy accountability to their child. In this scenario, the child eventually over-prepares and overreacts to the messiness imposed upon them.

An abusive innovation leader has little or no accountability to their employees. They exert toxic discipline and insincere empathy. Their employees over-prepare and overreact to a toxic leader. Abused followers analyze their options and act upon them earlier than what might otherwise be natural. Victims get out as soon as they can.

Good parents accept a certain amount of messiness from their children, but they prepare for it, react to it, and calibrate their discipline and forgiveness for the child. The parent's goal is to minimize messiness impacting family and friends. A good parent behaves with two-way accountability to their child.

A good innovation leader imitates a good parent in these ways. They have two-way accountability to their followers, and their goal is to minimize messiness hurting the value of the team and the outside stakeholder experience.

Keeping messiness levels low is valuable, but it also carries a risk. Like a helicopter parent who undermines their child's resilience, followers who always have a clean, easy path might be less ready to handle messiness on their own. Certainly, prepare the path for your followers, but also prepare your followers for the path. A good methodology prepares both.

Innovation Elegance keeps messiness low and prepares innovators for the path. Emphasizing documentation with Five Verbs reduces the messiness and poor accountability of meeting gridlock and verb sprawl. The Elegance methodology acknowledges the necessity and value of meetings and emails. But to minimize clutter, cost, and laboriousness, the Elegance methodology excludes meetings and emails from formal planning and status reporting. And when someone must change expectations about their contribution, it adds or removes their assignment among the Five Verbs.

To handle the inevitable messiness of teamwork, emulate the best culture traits of parenting. In its best form, parenting is about caring, not control. Parenting accepts all short-term messiness because the long-term payoffs are worth it. Parenting handles VUCA and responds with empathy and grace.

Self-sufficiency

Leadership is about making others better as a result of your presence and making sure that impact lasts in your absence.

~ Sheryl Sandberg (b. 1969), American business executive and philanthropist

Parenting repeatedly involves a sense of moving on. After some time, parents and children outgrow the daily watch and care. Children typically learn to fly, turning parents into so-called empty nesters. When the child succeeds on their own, they are self-sufficient. Innovation professionals face the same journey, growing from high dependence on their leader into self-sufficiency.

Sometimes children are dying to be on their own, and sometimes parents are trying to work themselves out of a job. One of the primary goals of parenthood is to raise an adult who can physically, psychologically, and financially support themselves and become a leader of their own family.

The goals of an empathetic innovation leader are the same—grow their followers, work themselves out of a job, and mentor the next generation to have their own leadership experience. Anything resembling possessiveness or loyalty stunts an employee's growth. Innovation repeatedly involves a sense of moving on.

Children learn from their parents for years through what parents say and do. Children decide what they will imitate and what they will do differently. Smart innovation professionals do the same thing as they mature: they observe what their leaders say and do. Ideally, they keep the good habits in their toolkit and discard the others.

Self-sufficiency is a phase of life that reconciles a mismatch common during the teenage years where they want the rights of adulthood without the responsibilities. People around them might tolerate this mismatch for a while, which avoids true self-sufficiency. Life is flexible about exactly when a child possesses both rights and responsibilities, but it's ruthless in that at some point, a child has both.

Some innovation professionals want the rights of leadership without the responsibilities. They want authority without accountability. This mismatch equates to power without leadership. Co-workers might tolerate this mismatch for a while, but instead of true self-sufficiency, it is mere selfishness, and a bad deal for their followers. Unbridled power undermines the self-sufficiency of followers. When intelligent, self-respecting followers observe this, they 'fly the nest' as quickly as possible.

Good parents are the ultimate servant-leaders. They take a child from beginner to self-sufficiency for their benefit, constantly investing in them. A good innovation leader is a servant to their employees—creating the culture, habits, and language for their followers to flourish and ultimately lead their own team. Companies are not obligated to care about their people, so companies that do have a competitive advantage.

Martial Arts

The secret of change is to focus all your energy not on fighting the old,
but on building the new.

~ Socrates (469-399 BCE), Ancient Greek Philosopher

M ost innovation work involves being shoulder to shoulder with people on the same team. But humans are naturally competitive, and some are wired to be so competitive that they prompted the terms 'hostile coworker' as well as 'toxic work environment.' These individuals are common enough that encountering them is inevitable. To avoid injury and salvage a morsel of value for your project, consider adopting some tools from the martial arts.

Habitually confrontational employees seem to always run contrary to the mission or people they see as a threat—earning them the label 'contrarians.' Contrarians create and circulate problems and rarely propose good-faith solutions. To minimize hostility and damage in your team, you need skills and tools to protect yourself, advance the work, and minimize harm to the team, including to the contrarian.

In combat, people get hurt. Physical confrontations can cause physical damage, and verbal confrontations can cause emotional and psychological damage. This is most pronounced in the victim, but in a conflict the attacker receives injury as well. One martial art, Aikido, distinguishes itself because of its unique goal to protect both the victim and the attacker from injury.

Aikido was developed by a Japanese man named Morihei Ueshiba. Early in life, Ueshiba studied and mastered numerous martial arts, but felt each was lacking. He created a genre for practitioners to defend themselves while protecting their attackers from injury. Aikido is not for fighting or defeating enemies but for joining people together in peace. Instead of martial arts training to conquer others, Ueshiba used it to refine and perfect the self. Aikido is a commitment to the peaceful resolution of conflict whenever possible.

Aikido is non-competitive. Traditionally, tournaments and matches are discouraged. Students do not progress by defeating opponents. They advance by demonstrating Aikido's philosophies and exercises.

The word Aikido comes from ai (harmony), ki (spirit, energy), and do (path, way). This roughly translates to 'the way to harmonize energy.' Successful Aikido is when the defender, the *tori*, blends and controls attacking energy, and the attacker, the *uke*, becomes calm and flexible in the disadvantaged, off-balance positions in which the *tori* places them. An innovation leader with a contrarian finds themselves in both roles—*tori* and *uke*—and so benefits from the skills of both.

Aikido is not easy. Defending yourself without injuring your opponent is arguably more difficult than defending yourself by injuring them, because you're effectively protecting two people. Although they treat you like an enemy, you treat them like a training partner. You operate with empathy and grace and will likely receive none in return. But this approach is more sustainable than abandoning empathy for your opponent.

Entry

It is not the critic who counts . . .
The credit belongs to the man who is actually in the arena.
~ Theodore Roosevelt (1858-1919), 26th President of the United States

In Aikido, confrontation is not a conventional competitive match. The interaction begins with the uke initiating an attack, called 'entry.' There are many ways in which the attack can begin: from in front or behind, with a limb (a strike), their body (grappling), or a weapon (knife).

An attack in a team environment also has many ways to begin. Contrarians undermine meetings by going off topic, monopolizing conversations, or through inconsistent attendance. They might undermine email correspondence by being ambiguous or involving the wrong stakeholders. The contrarian might undermine documentation by neglecting their assignment to Draft, Review, Revise, Approve, or Distribute an asset.

Of course, rehearsing Aikido in a dojo differs from being attacked by someone who means harm. For an innovation team, the first step in handling a contrarian is accepting that your project has a contrarian. Even experienced project managers sometimes are in disbelief that someone in the same organization seems to want failure and chaos. Every experienced project professional will tell you contrarians are common enough that most projects are simply going to have such a person.

Many people in power have a high tolerance level for internal conflict. Hostile coworkers are simply common. And conventional innovation methodologies address software, not people or hostile coworkers. Whether they intend harm or not, do not be surprised when a contrarian attacks you.

Instead of physical hostility, a contrarian is hostile toward harmony and collaboration. A contrarian often claims to be transparent, but don't confuse genuine transparency with belligerence and a tsunami of opinions. A contrarian likes chaos. Instead of a healthy agreement factory, a contrarian likes a disagreement factory. They prefer to work outside existing systems. They give discipline labels such as in the weeds, overkill, and over-analyzing. They are skilled at labeling others' ideas as overly complicated while complicating matters themselves in their own way. A contrarian considers Five Verbs both restrictive and excessively complicated.

Contrarians have a clear preference among the three communication channels of meetings, emails, and assets. Assets make them feel cornered. Emails leave a written record, and thus pose an accountability risk. Contrarians prefer meetings because meetings provide the most freedom with the least accountability.

In calculating ROI for innovation work, no team formally factors in the negative impact of contrarians, and thus they are doomed to underestimate how contrarians impact every dimension of teamwork. They hurt speed, quality, and create waste. They create new vigilance needs, increase variability, and keep a team off balance. Their negative impact is not a future risk—it is guaranteed and immediate. Their behavior impacts the morale of many people and does significant damage to the value proposition of their projects.

The driver of most contrarians is insecurity in their own value. They might have a zero-sum game mentality where they only sense success in others' failure. To feel more secure, they redirect team energy to their personal opinions, status, and goals. They especially draw attention away from coworkers whom they see as their biggest threat.

Many contrarians have a hero mentality where the only antidote for their insecurity is the opposite—the feeling of invincibility. To overcompensate for their insecurity, contrarians like to be seen as indispensable.

The most extreme versions of contrarian qualify for the most extreme labels, such as predator, infiltrator, or saboteur. Their energy is not on themselves; the most extreme contrarians intend to destroy entire projects, companies, and lives. They diminish others' value and morale with far-reaching effects.

Regardless of a contrarian's source of hostility, accept that contrarians are common. They attack you and your team-oriented colleagues. Their impact on culture is consequential and enduring. You might not know if they want you to fail, but you must manage them as if they do.

Blend

This is your world. I'm just in it.

~ Unattributed

When attacked in Aikido, your first step is to blend with your attacker. Blending meets what the *uke* propels at you (their arm, leg, or weapon) with your own hand, wrist, or arm. You must temporarily move in the same direction in order to divert that movement together as one body. Blending allows motion to continue without direct conflict. Ignoring or resisting the attack is prone to magnify the conflict.

Stepping entirely out of the way is also a temporary option, as it buys time to recognize what's happening. It can also put you in a more advantageous position. But when attacked in an innovation project, procrastinating engagement risks injury to yourself and your colleagues. Don't avoid the inevitable. Engage the attack and blend.

A contrarian's typical first attack is on a specific asset (e.g., a Project Charter or a Process Flow). The attack typically occurs in a Review and Revise meeting for that asset. The attack derails the conversation off topic and off balance.

Blending with the attack takes the form of documenting meeting minutes. The meeting minutes should explain why and how the agenda was disrupted and whether the asset is still in scope. This is blending—temporarily going in the same direction.

Blending conveys alignment with the opponent—being on the same team. This involves looking and moving in the same direction for a moment—a 'motion of convergence.' Harmonizing with your opponent gives them the illusion that they are getting what they want. This temporary alignment conveys to the contrarian that you are giving them the benefit of the doubt that they are acting in good faith. Of course, they may be acting in bad faith; you should blend either way.

Another way to blend is to adopt the contrarian's language. A contrarian dislikes the tidiness of Five Verbs. Be ready to embrace verb sprawl, or as the contrarian sees it: verb freedom. A contrarian dislikes active voice, so be prepared to talk and write in passive voice. And a contrarian dislikes a formal, unambiguous project plan, so be ready to reinvent the wheel with a fluid task list with few dependencies, assignments, or milestones.

Admittedly, blending with the opponent of your innovation team steals attention away from focused and sincere collaboration among team members. Blending becomes the bottleneck for the project, but it is just the start of engaging the opponent.

Control

> *Being able to admit when you're wrong is important,*
> *but so is standing up for yourself when you're right.*
>
> ~ Suzy Kassem (b. 1975), American poet, writer, and philosopher

The ideal scenario in Aikido is that the defender, the *tori*, maintains control of the situation and gains a tactical advantage over the attacking *uke*. An advantage in Aikido is specifically to take what's extended to you, pull your opponent in an aligned direction, and rotate your shared focus. The tori never allows an opponent to relax, but instead continually redirects their power.

The ideal scenario in innovation is that the methodology maintains control of the situation and gains a tactical advantage over the contrarian. To gain an advantage, take what's extended to you, pull it in an aligned direction, and rotate your shared focus. The contrarian has extended to you some combination of five things: meeting minutes, a meeting factory, verb sprawl, passive voice, and a task list.[12] Control these together with the contrarian and rotate your shared focus.

Regardless of what's best for your team, blend with the contrarian. Seize what they give you, stay within their comfort zone, and adopt their preferred channel. Groom the illusion for the contrarian that they are getting what they want.

---

12 Without the contrarian, these would instead be an asset portfolio, an agreement factory, Five Verbs, active voice, and a project plan.

Throw

Every absurdity has a champion to defend it.

~ Oliver Goldsmith (1728-1774), Irish playwright and poet

Once the *tori* has control, one option is to throw the *uke* to the ground. A throw can be effective in halting the *uke's* aggression as the separation from the *tori* gives grace to the *uke* in the form of time to think about what to do next and whether to risk getting thrown again.

Typical throws involve the tori administering a joint lock—manipulating a vulnerable joint to its maximum extent of motion. The uke can only ease the pain by having the rest of their body follow the pressure, and their momentum typically sends them to the ground.

An innovation professional's version of a throw is to exaggerate the contrarian's momentum. Instead of being vulnerable at a joint, a contrarian is vulnerable—accountable—to their sponsor or the whole team. The 'manipulation' of the sponsor or the team is their receipt of the exaggerated communication momentum, to the point of absurdity. For example, instead of a portfolio of structured assets, absurdity is a portfolio of meeting minutes. Instead of Five Verbs, absurdity leans into verb sprawl. These are painful for everyone. The contrarian can only ease the pain by following this pressure and answering directly to the sponsor or team.

If a contrarian attacks once, they might be acting in good faith. It could qualify as a sampling error. If a contrarian attacks repeatedly, they are acting in bad faith. Their behavior is purposeful, and thus systematic. Systematic errors are sabotage. Not only is the contrarian not accountable to those they attack, but they also convey that they want to injure them.

A contrarian mindfully propels disruptions at the colleagues they want to injure and not at others. To keep the targets safe, locate the people to whom the contrarian will be accountable (e.g., their manager or customer). If you need escalation beyond that, consider your formal whistleblower options. Escalation increases the risk of injury to the contrarian, but their persistence in attacking means the fight is worth it to them. Escalation signals that you don't want to injure the contrarian, but you *would* like some help to neutralize them. This is like attempting to throw the *uke* in a different direction after the first doesn't work.

Exaggerating your opponent's momentum gets labels like passive-aggressive and vicious compliance. That may be true but consider which is the bigger sin. *Toris* don't start fights; they finish them. Exaggeration resembles what ancient Greek philosophers called *reductio ad absurdum*, which asserts that your opponent's logic leads to absurdity. For innovation work, absurdity is meeting gridlock, pausing a project, or declaring it a failure.

Everyone becomes off-balance when you exaggerate the contrarian's communication momentum (e.g., creating a meeting factory). The goal is for the sponsor or other team members, separate from you, to ask the contrarian to stop their absurdity. When you throw a contrarian, you give them the grace to get back up, think about what to do next, and whether it's worth getting thrown again.

Immobilize

> *I am here today to cross the swamp, not to fight all the alligators.*
> ~ Rosamund Stone Zander (b. 1942),
> author and corporate leadership coach

The second option for a *tori* in control is to administer an immobilization technique. Immobilizing the *uke* is a less attractive option since the *tori* must stay connected to the *uke* to apply pressure, and the *uke* experiences pain longer. Disengaging from this kind of confrontation is more awkward.

Whether the *tori* throws or immobilizes mostly depends on the *uke*, their momentum, and their willingness to continue moving. If the *uke* is unwilling to be thrown, the *tori* changes their lock until the *uke* is on the floor, face down and immobile.

In an innovation setting, two forms of immobilization exist. If the contrarian wants to avoid having their momentum used against them, they might decide to stop moving (contributing) or encourage others to not contribute either. They might avoid meetings, emails, and contributing to documentation. You likely have no control over their meeting attendance or email responses, but if you control the project plan,[13] you can remove their assignments and be transparent about it. Informally, they might continue being part of the project, but formally, they don't contribute to project assets, effectively removing them from the project.

The second immobilization option is to pause or cancel the project. It's bold and dramatic, but this 'stops the bleeding' for the project. Pausing is reversible, and preferable over an irreversible waste of time and talent when a project's value proposition and employee experience crumble.

Many project sponsors have allowed lousy projects to hobble along due to a contrarian's input, and they often wish they had paused or canceled the projects instead.

Calm

Please put on your oxygen mask before helping others.

~ flight attendants everywhere

For an innovation professional, the harsh reality is that the contrarian is often in control. You must learn how to land when the contrarian maintains the upper hand. In this case, you need the tools of the *uke*. In the *uke's* role, you are at a disadvantage. You are off balance and vulnerable, so focus on regaining your balance and covering your vulnerabilities.

---

13 Assumption: you control the contrarian for purposes of the project, but not for purposes of their employment.

The first tool is staying calm. Do not be surprised when attacked; instead, study your attacker so that you can respond to the opportunities that present themselves.

For an innovation professional, the easiest way to stay calm is to be empathetic toward a contrarian. A contrarian's motivation is insecurity: something about the project threatens them. Innovation competes with the status quo, and eventually change replaces stillness. It's rational for a contrarian to know that their status, financial security, or job security are vulnerable. A contrarian's role in the project makes them feel insecure or trapped, and they react with contempt. Once you refuse to be surprised and instead empathize with the contrarian's perspective, staying calm is easier.

Calmness with a contrarian means executing the Five Verbs framework for as long as possible, fully aware they threaten contrarians who want to throw you and the project off-balance.

Flexible

If you're going through hell, keep going!
~ Winston Churchill (1874-1965)

In the dojo, Aikido practitioners spend approximately half their time acting as *tori* and half as *uke*. Although the *uke* is the original attacker, the tables turn if the *tori* successfully blends and gains control. To avoid injury, the *uke* yields and becomes flexible. A successful *uke* gracefully receives the lock or throw from the *tori*, follows through on the motion, and regains balance.

The ideal *uke* acknowledges they do not control the action. Rather than rigidly continuing to resist, they are elastic to the *tori*, granting them this small success. This allows them to escape harm for themselves, and safely and gracefully regain their balance while they decide what they want to do next.

A skilled innovation professional minimizes injury to themselves and their attacker by allowing the contrarian some degree of success. Allowing success and avoiding the extremes of rigidity requires tools that give you flexibility while neither you nor the contrarian are clearly in control. This corresponds to blending (although this blending is on the contrarian's terms). This gives the contrarian the illusion of getting what they want.

In a team environment, indefinite blending is laborious, since it surrenders and leans into meeting gridlock, email overload, and counterproductive documentation. Uncollaborative blending leads to fatigue for all team members as well as a slow demise of the team culture and the project's value proposition. Blending only stops when you or the contrarian gain control, or another team member intervenes and takes control.

When a contrarian gains control, sometimes your best option is merely surviving and not getting fired. This does not mean just accommodating the contrarian's advantages. Staying flexible is strengthening your own options, such as enhancing career security while the contrarian undermines your job security. If you're fired, you will be of no help to anyone in the organization. Signal widely that the contrarian is in control and keep moving with them in harmony.

Rollout

If at first you don't succeed, try, try again. Then quit.
No use being a damn fool about it.

~ W.C. Fields (1880-1946), comedian, actor, and writer

In a conflict, the goal is to avoid injury to yourself and your opponent. Avoiding injury requires separation, which can't happen when immobilized. Separation requires falling gracefully and then rolling away from your attacker.

Rolling with grace requires practice. It's important to relax as you roll and get low whenever possible. The ground is your friend, since a shorter distance reduces the risk of injury and improves your angle to get back up again.

Rolling away from a contrarian in an innovation setting takes a few forms. The most dramatic is to exercise your career security at the expense of your job security and quit the project or organization. It's a shame to leave an organization because of a bad apple with a sponsor, but options are attractive, and exercising them is an adventure. Options are advantages that insecure contrarians don't have. Especially if the contrarian's behavior is unethical, avoid being complicit. Fall quickly, gracefully, and roll out of the organization.

The Myers-Briggs Type Indicator (MBTI) personality framework is a much less dramatic tool to diffuse conflict. At the time of writing this book, the MBTI framework is falling out of favor, since it suggests a person only operates within a single profile. Even so, the MBTI framework helps explain conflicting personalities and behaviors.

Some personality types embrace big-picture concepts and dismiss detail as 'in the weeds,' yet cultivate ambiguity themselves. Team members who overcommunicate might feel that others hide information or have nothing to say, while the under communicators might be waiting for their opposites to stop monopolizing the dialogue. Some personalities prefer structure, while others prefer fluidity. MBTI is a safe framework to stop employees from demonizing each other and instead, attribute team dysfunction to work styles of incompatible personalities.

Another tool to defuse confrontation with dignity is 'Thank You Because' (TYB). Conceived at The Second City Improvisational Theater, TYB is useful when disagreement is strong, and both parties want the relationship to survive. TYB provides a bridge to keep the conversation and the relationship going. Disagreeing project team members might say to each other, "Thank you for explaining because I know you want to do quality work, keep things moving, and report positive news." TYB is a great tool to diffuse a confrontation, acknowledge differences, and keep everyone in the conversation.

A job is not worth one's sanity, reputation, and self-respect. Although a contrarian won't let go of their ego, let go of yours. The world has enough sore losers. In Aikido, the *uke* is a good loser—they're calm, flexible, and roll out of confrontations. MBTI and TYB help you yield to a contrarian, roll out of conflict, and regain your balance without leaving the organization.

Contrarians are expensive. They disrupt your innovation methodology and increase the cost of innovation. Being unprepared for contrarians multiplies costs. While the tools of Aikido seem paranoid and disruptive in their own way, they mitigate runaway costs by neutralizing the contrarian, euthanizing the project, or both.

When an innovation team has a contrarian, the martial art of Aikido is practical, elegant, and empathetic. It acknowledges insecure employees and prevents a toxic culture from forming. The art's principles minimize injury, provide a graceful exit from conflict, and minimize erosion of the value proposition.

Improv

Good improv is all about listening, reacting in the moment, creating, and supporting the ideas of others.

~ Tom Yorton (b. 1966), co-author of Yes, And and former CEO of Second City Works

Some business activities are tightly controlled—they are scripted and even automated. Many are not. Improvisation is the art that helps businesses execute intelligently when left without a script.

Working without a script doesn't mean surrendering to paralysis or a chaotic free for all. Improv encourages you to work without knowing where it will lead. Improv keeps work flowing and gives your team the confidence to proceed in safe, fun, and promising ways.

It's common for a team to fall casually into a conversation that resembles an improv session. This conversation has everyone contributing with small additions. The conversation is full of acceptance for its twists and turns.

Improv is uniquely valuable when the team lacks clarity about the best direction for the conversation. Such a team needs to explore, get everyone involved, proceed with ideas that catch on, and ignore ideas that don't. Improv is a great tool when a team worries its output is mundane and lacks ingenuity. Improv encourages entrepreneurial thinking and is an excellent tool for discovering positive surprises.

The mother of improvisation is a woman named Viola Spolin. Viola's early career in Chicago was doing social work. During the Great Depression, she formulated tools and games to stimulate creative expression and social behavior among inner-city and immigrant children. Eventually, she took these games to the stage as an acting coach.

Through the games, players learn specific skills such as reducing self-consciousness, pre-planning, and judgment. In this playful mode, the players are uniquely *present*. They produce instinctive, intuitive, and inspired choices spontaneously. Improv exercises are frameworks designed to fool spontaneity into being. In this game, spontaneity deserves a stage, and Viola wanted to impress upon her students that everyone is stage worthy.

People can be self-conscious when their culture glorifies exclusivity, individualism, and hero worship. Improvisation aims for the opposite. Improv celebrates everyone's contributions, the power of diversity, and thinking on your feet. Improv tools enable team members to direct their attention to the activity itself.

Employees need tools to get out of their own heads. The focus is on the ensemble experience, the employee experience, and the customer experience. As Scott Adsit, an American actor, comedian, and writer, once remarked, "All great improvisers are out to protect everybody else on stage and make sure they look good."

Improv's reputation is that it is a silly free for all, but that's not the case. Improv is a collaboration that guides all members to contribute repeatedly in small steps. Improv *is* fun and games, but the skills learned profoundly impact team output, employee engagement, and stakeholder impact. Improv skills improve teamwork by emphasizing listening, inclusivity, and authenticity.

The improv journey includes eliminating preconceived notions of what your results will be, building upon and encouraging diversity in what your team members give you.

Theater for a VUCA World

Dreams the way we planned 'em, if we work in tandem.
~ Elphaba to Glinda in the musical *Wicked*

Improv is theater where, instead of executing with a literal script, you perform with an abstract script of rules. Improv rules encourage the story to unfold spontaneously without a firm plan or previously known result. The team makes something out of nothing, and every actor contributes to the resulting body of work.

Improv is not impromptu ad-libbing; it has rules and requires practice. Instead of practicing a literal script, improvisers practice the rules. The output of conventional theater is not spontaneous, but the output of improv is, encouraging you to think on your feet.

You build improv skills by participating in improv games. In playing the games, you recognize patterns, set boundaries, and achieve rhythm. Improv teaches that obstacles are gifts and that you do not have to tackle obstacles alone. It leverages the strengths of individuals through the diversity of thought each person brings.

Once you embrace the experience of improv outside your work, you will embrace the experience inside your work. It will improve teamwork and team results. An improvisational conversation's unplanned and poised rhythm resembles a graceful factory.

The business world increasingly embraces the notion that diversity, in many forms, is a competitive advantage for companies. Improv puts diversity to good use. Diverse voices that span demographics lead to diverse ideas, which uncover positive surprises. Positive surprises are a competitive advantage.

Businesses can't guarantee where they will end up; therefore, business is one big act of improvisation. Rules of improv ease innovation because the value of preparation is limited. When you face a scenario needing improvisation, it is counterproductive to over-prepare because you do not know precisely what is coming next. Improv skills help you harness a VUCA world. They are the perfect tool to anticipate and welcome surprises.

"Yes, And"

It's OK if others share our ideas as long as they build upon them.
It's called progress.
~ Simon Sinek

When you put improv into action, your first words are, "Yes, and." "Yes, and" is a sequence of one actor offering an idea, other actors affirming it, then building onto it with something of their own. Each actor is called upon to 'explore and heighten.' It looks like this:

1. It's hot outside.

2. Yes, and when it's hot outside, I want to go to the beach.

3. Yes, and when I go to the beach, I hope the waves are big!

4. Yes, and when the waves are big, I have to be careful.

5. Yes, and when I'm careful, I won't get sunburned.

6. Yes, and when I'm sunburned, I regret going to the beach!

Martin De Maat, the past artistic director at The Second City Improvisational Theater, once observed, "The fun is always on the other side of Yes." The method gives every idea the chance to proceed and flourish.

In business, like in improv, you do not have to act on every idea, but you do have to give every idea a chance to be acted on. You do not have to love every idea, but it doesn't hurt to appreciate it for just a little while.

Bad ideas can simply be a bridge to better ones. Improv encourages you to contemplate potentially useful but incomplete ideas before they get shut down. Hidden gems are buried in there somewhere.

Ensembles fail together, so always take care of your partners. People are more willing to contribute fully if they know the risk is shared among the many. Or as Tina Fey, American actress, comedian, and playwright, remarked, "There are no mistakes, only opportunities."

Improv exercises often roleplay with "Yes, but" and "No." These games feel different than a game of "Yes, and." These phrases are scene killers and bring conversations to a dead end. They see alignment as an enemy.

"Yes, but," and "No" appear in the workplace when one person wants to control the dialogue. That person might want to steal focus or be proven right. These are enemies of high-functioning ensembles. "Yes, but" and "No" often occur when someone is more interested in showing off their brain power than solving the problem. When dominated by those who never surrender the need to be right, every disappointment is magnified, and decisions are met with hostility. Improv is a threat to anyone who wants to monopolize power and attention.

The asset portfolio is the written form of "Yes, and." Assets don't say "No" to upstream assets. Every asset 'explores and heightens' because it builds upon some combination of upstream assets. "Yes, and" reinforces the notion of traceability and prevents orphaned ideas in an innovation team's work.

Co-creation

None of us is as smart as all of us.
~ Ken Blanchard (b. 1939), author of *The One Minute Manager*

Even brilliant people do not accomplish great things in isolation. Everything in life worth creating requires some form of team. This is co-creation.

Co-creation is critical for innovation teams to move past the outdated cliché: a team is as strong as its weakest member. Instead, the improv world teaches: a team is only as good as its ability to compensate for its weakest member. A great ensemble contains a variety of individuals with strengths that will be enhanced and whose weaknesses will be minimized by the group dynamic.

Innovation teams require diverse skills. At different points in a project, every team member is the weakest member of your team. You want team members to step up into their expertise and elevate their colleagues instead of focusing on someone's weakness. Co-creation shapes a culture of 'got your back' instead of 'watch your back.'

In co-creation, autonomy is low; you are never on your own. To emphasize a sense of togetherness, the improv profession uses the French word for together—ensemble—instead of the word team, which acknowledges competition. In innovation, no piece of work is built in isolation.

Take the example of Five Verbs. Assigning one person to all of them makes no sense. And while co-creation is happening, there is no sense to have competition—another team. There is only a sense of collaboration—your ensemble.

Co-creation doesn't need much to succeed; it requires a host and a collaboration space. A good host creates a space less like a boxing ring and more like a campfire. When team members exchange ideas over a campfire and give up the notion that they have to be the most intelligent person in the room, the magic of co-creation can happen. The Elegance methodology emphasizes documentation so strongly because documentation is the campfire—the collaboration space. What constitutes success for co-creation is simply that the ensemble builds something together.

In any ensemble, every member wants to be heard. Co-creation fosters sustainable engagement and positive morale from all members of the ensemble. It requires that each person be committed to each other's safety, morale, and contribution.

None of us is as smart as all of us. Improv provides the arena—the campfire—for the ensemble to co-create.

Follow the Follower

The first follower is what transforms a lone nut into a leader.

~ Derek Sivers (b. 1969), American writer, entrepreneur, and musician

Everyone has some form of genius. Everyone has a passion they can bring to their work. Mature leaders know they should not always be the one leading, per se. Improv nurtures an environment to *shift leadership*—for leaders to follow the follower (FTF).

An improv ensemble continually rotates leadership and followership roles. The ensemble manager knows when to lead, follow, and get out of the way. In innovation, a mature leader does the same. They are mindful of when to be hands on versus hands off—when to intervene versus when to let junior employees run with things (i.e., FTF).

Although an FTF culture is healthy for long-term career development and security, it's also valuable for short-term value. Any business likes to not 'leave money on the table.' An FTF ensemble doesn't leave talent on the table. Instead of individuals monopolizing as they exercise their talent, an FTF ensemble democratizes talents.

FTF unleashes talent—brainpower—across the team as well as proactive potential in individual employees. It undermines bias in thinking, silos in behavior, and bottlenecks in progress. Individuals thoughtfully alternating between leading and following is a symptom of a working team.

FTF brings out the best in thinking and doing across the team. It fully uses talent for short and long-term goals, weaving that talent in and out of workstreams. Talent weaving encourages any member of your ensemble to assume leadership for as long as their expertise is needed. In an encouraging dilemma about which is first, weaving talent in and out builds a high-functioning ensemble, and only in a high-functioning ensemble can you continually weave talent.

In an innovation team, the evidence of talent weaving resides in the project plan. When you add and inspect assignments in a project plan, ensure a variety of people are assigned to the Five Verbs process. Avoid the monotony of someone's name appearing too much with one verb. It leads to monotony, fatigue, and bottlenecks. Aim for as much balance as expertise and learning allow.

An intelligent follower distinguishes between respect and reverence. Someone who has been a follower for a reasonable amount of time has observed the current leader and contemplated how they would lead differently. A fresh leader might know what legacy leadership reveres as sacred and untouchable. Respect is great; however, reverence can be the enemy of change. A new leader respects without over-reverence.

A culture of FTF is straightforward because it doesn't require any preparation. It's valuable because it optimizes engagement, enthusiasm, and energy.

Bring A Brick, Not A Cathedral

The river swells with the contribution of the small streams.

~ Proverb of the Bateke people, Republic of the Congo

Some people do too little for their ensemble. Some people do too much. The right amount is what the improv world calls, "Bring a brick, not a cathedral."

The best improvisers know that the ensemble, working together, ultimately creates the full scene. Individuals are responsible for bringing the next idea or piece of information, that is, a brick. Your ideas are critical, not because they are brilliant, but because they encourage the flow of ideas from others. Every idea builds upon the previous and contributes to the next.

When you set the foundation, you make the job of others easy to contribute to the full 'cathedral.'

The field of neuroscience teaches the closure principle—that human beings are wired to contribute to incomplete ideas to make them whole. We tend to respond positively when invited to contribute to a concept that is not yet fully formed. But a closed environment, dominated by those who need to be right, alienates others, relegating them to spectators.

For an improv director, one sign of success is balance in the show: actors contributing approximately equally. The director tries to assign equal amounts of speaking and non-speaking parts. An improv director also monitors the audience's and actors' reactions for authentic engagement and enjoyment, determining what works and what doesn't.

An attentive innovation leader also aims for such balance—similar workloads, contribution levels, and balanced assignments across Five Verbs. An attentive leader monitors what works (bringing a brick) and what doesn't work (bringing a cathedral). They monitor reactions for authentic engagement from all stakeholders.

As technology continues to take ownership of repetitive tasks in the business world, the future of human labor requires you to leverage what makes you most human. The jobs of the future will rely increasingly on divergent thinking—a kind of vigilance that robots cannot do. Kelly Leonard is right when he says, "Robots cannot improvise, and they're not funny." The art and tools of improvisation are instrumental for a culture of empathy, entrepreneurship, and exploring your innovation frontier.

Music

I love the humanity to see the faces of real people devoting themselves to a piece of music. I like the teamwork. It makes me feel optimistic about the human race when I see them cooperating like that.

~ Sir Paul McCartney (b. 1942), singer, songwriter for the Beatles

A reasonable expectation of an innovation team is that they will support each other and collaborate toward common goals. But friction, franticness, and failure rates expose the reality that many innovation teams have mediocre support for this level of collaboration. Making music—specifically a symphony—is an excellent model for an innovation team to achieve basic teamwork expectations.

Engaging stakeholders late in the project is a common mistake for innovation teams. Innovation professionals make the wrong assumptions, work in silos, or get lazy about the full impact of some change initiatives.

Another common culture trait is extremes in speed. Some organizations try to do too much, causing franticness or traffic jams that bring work to a crawl. And many cultures accept clashes of various kinds, such as personalities, schedules, and vision. Many innovation teams hold on to negativity about a project. These culture traits are anything but pleasant to hear about or listen to.

Executing like a symphony solves these problems. At a symphony's inception, every member and stakeholder knows the ensemble needs diversity to create a rich experience. A symphony has variety in the composers it chooses, the instruments it uses, and even the volume and speed at which it plays. A symphony paces itself—tightly governing the schedule in which it learns the music for an upcoming concert and the speed *within* every piece of music.

It's not an exaggeration to say a symphony prohibits clashes. Expectations are so clear—with such low ambiguity—that confusion and conflicts don't make sense. Low ambiguity breeds early and sustainable enthusiasm for collaboration and high confidence in the prospects for success. Confidence and enthusiasm feed a culture of positivity.

Diversity, pacing, harmony, and positivity aren't valuable for the sake of themselves. They're valuable because they create a pleasant experience for the audience and the musicians. Monotony, franticness, stagnation, and negativity are unattractive culture traits that reduce the value of the collaboration. Making music is customer-centric: people want to watch a symphony. Many would like to be part of it!

Because innovation teams struggle with culture traits like diversity, pacing, harmony, and negativity, they can benefit from a symphony's most basic culture traits.

Diversity

A symphony must be like the world.
It must contain everything.

~ Gustav Mahler (1860-1911), Austrian composer and conductor

A symphony incorporates a variety of instruments to create a rich experience for its audience and its musicians.[14] Innovation teams need diversity to

---

14 It is important to note that symphonies still lag behind other fields in the diversity of their personnel.

create rich experiences for their customers and employees. Diversity takes the form of representation, project selection, and intensity.

A symphony must be thoughtful of its instrumentation—the combination of strings, brass, and woodwinds that comprise the music. A symphony must also be mindful of competence level so every musician can learn their part. If a symphony lacks the right personnel, it becomes risky for dozens of people to spend time on a mediocre outcome or unachievable goal. Together, instrumentation and competence ensure the symphony has the proper representation for the collaboration to succeed. The ramifications of rehearsal and performance schedule delays are worth getting the audience experience right the first time.

Healthy innovation teams do a similar representation check. If a skill or a stakeholder group needs to be more represented, a diligent project sponsor secures the right employees to join the innovation ensemble. If a team member's competence level hinders them from contributing, they must be trained to fulfill their role and not obstruct the project's speed and quality. When an innovation team lacks the right personnel, proceeding without them propels the project into errors, a mediocre outcome, or an unachievable goal. Delays in the project and go-live event are worth getting the customer experience right the first time.

Another aspect of diversity is the music itself. The director selects a mixture of music to bundle into an upcoming concert. They prioritize a variety they believe interests the musicians and pleases the audience. They consider recent shows, the time of year, the mood of the community, and what might belong in subsequent concerts.

Preparing a concert is a project, and preparing an innovation go-live event is a project. Every innovation project is derived from a queue of people, process, and technology changes. Every project is either a bundle of these changes or one big change.

The projects are prioritized and reflected on a roadmap. The roadmap reflects the diversity of innovation work. A roadmap can change to reflect shifting priorities and new information.

Music uses changes in volume—called dynamics—to keep the music interesting. Playing or listening to music that has a singular volume is exhausting to play and monotonous to hear. Dynamics are also desirable for innovation team members and their customers. Customer and employee experiences that are always intense and high volume are exhausting, and long periods of dull activity are monotonous. A mix of intense and relaxed routines is typically the ideal experience for innovators and their audience.

Symphonies can include solos, where individual instruments have moments to shine, but individual stardom is never the goal. Rather, diversity is a primary goal of a symphony—showing off different instruments, composers, speeds, volume, and harmonies. Members join to be a part of something bigger than themselves.

An innovation ensemble is richer when it includes as many voices as feasible. Each voice can still maintain its identity. A clever analogy about this diversity says a team is less of a melting pot and more like a tossed salad. Marginalizing ingredients makes the salad mundane. Adding ingredients adds richness. Diversity creates richness in your innovation ensemble.

Pacing

The best way to learn is through the powerful force of rhythm.

~ Wolfgang Amadeus Mozart (1756-1791), Austrian composer

Another critical element in music is the rehearsal schedule. The schedule synchronizes all the moving parts to create one integrated *performer* experience. The rehearsal schedule shows interim milestones to ensure the ensemble paces itself well as it works toward the concert date.

The innovation equivalent is the project plan. It synchronizes all the moving parts to create one integrated *employee* experience. The project plan shows interim milestones to ensure the team paces itself well, working toward the go-live event.

A music conductor sets the tempo—the speed—of the music. Although there are typical speeds the composer recommends, the conductor might choose a slightly faster or slower tempo. The conductor might even change the tempo in the middle of the song.

Leading the ensemble extremely fast or slow is counterproductive. The point of the music is to spend time in that texture of teamwork; it makes no sense to be in a hurry. Music gives you space to sit and take the time to recognize the variety of what's going on around you. However, there is no downtime, either. The ensemble waits for no one. Within a song, the tempo ruthlessly moves on.

An innovation leader sets the tempo for their team. The speed can change throughout the project, but extremes undermine the value proposition. A frantic project invites chaos and burnout. Franticness undermines thoughtfulness, texture, and quality of teamwork. Meanwhile, a plodding project bores employees and aggravates customers.

Every project needs healthy, sustainable momentum. A healthy project has a simple and visible schedule that shouldn't have to wait for anyone. In innovation and music, every performer needs to keep up and contribute at a unified tempo.

Another counterproductive trait is 'cutting corners.' Undisciplined innovation leaders are known to shave off a fraction of a project schedule, which requires truncating code, testing, or training work. This ensures a negative surprise for the customer, later incurring costs higher than the original shavings.

A symphony does not cut corners and trim a small fraction of the music: it rehearses and performs the entire composition. Innovation teams should too.

Lastly, being synchronized matters in music and innovation. Disregarding synchronization undermines the team's speed. One person out of sync can cascade to undermine others who do synchronize. Healthy ensembles take pride in their tight synchronization, which requires as much effort being spent in listening as it does in making sound.

Synchronization in innovation conveys respect for everyone's time. Synchronization is a sign of respect to teammates and reverence for the music—the methodology—as written.

Harmony

Harmony makes small things grow; lack of it makes great things decay.
~ Sallust (86-35 BCE), Roman historian and politician

For a symphony, the composer documents the melody and harmony through sheet music. Sheet music sets boundaries for each instrument or voice, ensuring that performers 'sing from the same hymnal.' The sheet music is durable for future performers' reuse. Every performer is transparently accountable.

The idea of not having the music on paper and not having a standardized approach to documenting music is a recipe for chaos. Harmony in the music contains many moving parts, but nothing is wasteful or extraneous. Rehearsing music synchronizes all the moving parts to create one, integrated performer experience. The concert synchronizes all the moving parts to create one integrated *audience* experience.

For a healthy innovation team, collaboration is also documented—as assets. Harmony is present when a few assets are thoughtfully in motion simultaneously. The assets ensure employees are on the same page and have healthy boundaries. The assets are durable; present and future employees use them to ensure their transparent accountability.

The idea of not having processes on paper and not having a standardized approach to documenting the processes is likewise a recipe for chaos. Teams work to minimize extraneous and wasteful content. Project work synchronizes all the moving parts to create one, integrated employee experience. The innovation outcome synchronizes all the moving parts to create one integrated *customer* experience.

Periodically during a rehearsal and between songs of a performance, an ensemble tunes their instruments among themselves. Performers must tune to continually sound harmonious. In doing so, players adjust to each other to play at the same audio wavelength. While tuning, players adjust where they stand to avoid being in each other's way when watching the conductor. Auditory and spatial tuning are vital for performers to play well together.

A healthy innovation team also figuratively tunes its instruments to be on the same wavelength. Wavelength translates to the rules of engagement and methodology. An innovation methodology tunes your innovation ensemble by ensuring everyone is in alignment. The Elegance methodology contains a few dozen tools to manage and mentor your innovation ensemble into alignment.

Ensemble music minimizes the monotony of one voice or one instrument. Tension exists by way of dissonance. Alternating it with consonance—resolving the chord—is pleasant. Voices complement each other.

Healthy innovation teams likewise have little monotone, aiming instead for harmony. Tension—divergent thinking—exists. Alternating it with convergent thinking by way of resolution and optimizing globally is pleasant. Team members respect boundaries and contribute differently, allowing their skills and perspectives to complement each other.

Harmony is a primary trait that conveys the quality of the team, whether in a symphony or innovation. Through harmony, the team creates moments that matter. Harmony is what makes the group an ensemble.

In music and innovation, harmony requires humility because such teamwork is never about you, an individual. The work is always about the team and its blend. Stardom and individual accomplishments don't matter to the customer; the value to the customer depends on what the *team* can produce. And that requires constant harmony.

Music—and many arts—are often labeled creative. This is more false than true. A composer is indeed creative. A composer thinks of a new way to assemble different instruments into a pleasant combination. But everyone else is just executing a script—playing the music as written—to fulfill the harmony the composer has in mind. Most of a musician's energy is not creative. It requires humility for a musician to have their creativity play second fiddle (pun intended) to harmony.

Every innovation team needs harmony. A symphony models synchronization, tuning, tension, resolution, and humility: all ingredients for rich harmony.

Positivity

The key to someone's heart is hidden in their playlist.

~ Unattributed

A common problem for innovation teams is when a decisionmaker sets expectations that feel irresponsibly high or low. This damages morale and shapes a culture of negativity.

An innovation decision maker who sets irresponsibly high expectations for their team typically leads to a high volume and quality of work for too little time and effort. The opposite scenario is launching a project that feels trivial, petty, or overblown. In this case, the complexity or volume of work is low. In both cases, the decision maker needs to be more in touch with their team and the nature of innovation work.

Changing scope near a go-live event typically has higher consequences for teams or customers than for the decision maker, whose post-project consequences are typically low. Irresponsible, impulsive decision makers might lack the incentive to grasp their team's reality, which breeds negativity in the team.

A musical director has incentives that prevent them from making such a mistake. These motivations foster positive morale and a culture of positivity.

As a concert approaches, the director can still remove a piece from scope if they feel it won't be performance-ready in time, say if a few senior performers have raised concerns. The consequences of pre-concert changes are low. But if a stubborn director kept an under-prepared piece of music in a concert, the resulting poor performance would likely lose the director their job, meaning the consequences for stubbornness are high. Decision makers have a high incentive to avoid irresponsible expectations of the symphony, which breeds positivity in the ensemble.

A contrast to a symphony, there is another form of music ensemble, a jam session. A jam session has a few different expectations than a symphony does; it has no sheet music, minimal expectations, and some surprises. It's reactive, experimental, and one-of-a-kind fun for small audiences—possibly the performers only.

The innovation version of a jam session is called a hackathon. Any success is accidental. Most innovation is not a hackathon because success should be realistic and expected, like a symphony. A jam session and hackathon have low formality, low focus, and low expectations.

One secret of music making is that the best seat in the house is not in the audience—it's inside the ensemble. The best place is being on the journey with talented players. If you enjoy teamwork, this secret applies to innovation ensembles too. The best seat in the house is within the team, surrounded by talented players on your innovation journey.

The culture of making music is a model for the culture of innovation. Both require diversity, a sustainable tempo, harmony, and a positive dynamic. Making music encourages responsible expectation setting. A deeply empathetic innovation experience resembles the music making experience, infusing synchronization and harmony among diverse musicians and audiences.

Dance

Leaders must encourage their organizations to dance to forms of music yet to be heard.

~ Warren Bennis (1925-2014), American author and pioneer of Leadership Studies

Innovation professionals, especially leaders, receive the benefit of the doubt for a few things: their self-awareness, their self-management, their ability to partner, and their good stewardship of their organization. But these are common problems for an innovation team.

While most innovation professionals have the technical competence to step into a team environment, some need a more accurate grasp of their strengths and weaknesses. Once they have a strong sense of self-awareness and self-management, they might be a great individual contributor, but not a great partner or collaborator. And finally, even if someone is an adequate partner, they might not rise to care for their innovation community.

The art of partner dance provides a model for innovation teams to build these culture traits, allowing them to shape positive experiences for employees and customers. Beginners keep their learning and attention on themselves and on aspects of dance like mechanics, retention, and goals. As a dancer masters these, they work harder at being who their partners need them to be while respecting boundaries, watching for blind spots, and recovering from mistakes. Mature dancers are stewards in the dance community while rotating partners, making dance less terrifying for beginners, and embracing that learning never stops.

Of all the empathetic arts, dance is most vulnerable to rejection by innovation professionals. Some don't enjoy participating or being in an audience, so they reject the lessons to be found there. But even if you don't enjoy dance or take it seriously, don't put yourself at a disadvantage. The culture traits of partner dance are great teachers for competition and collaboration, and they provide unique advantages.

Partner dancing is among humans' most collaborative, empathetic, and elegant activities. It's also the most exposed, intimate, and vulnerable. It's easy to get intimidated and quit. Dance requires humility and perseverance to get through being a beginner and arrive at a place of confidence. But once you become a steward of the art, you can infuse empathy and grace into everything you do. The art of dance provides culture traits for innovation teams to learn, persevere, and profit.

Self-Management

You can't really be present for the people in your life
if you aren't taking care of yourself.

~ Kerry Washington (b. 1977), American actor, producer, and director

Before a dancer worries about a partner, they need to worry about themselves. Before they can enjoy dancing with someone else, they must enjoy dancing on their own. Before you join an innovation team, make sure you manage yourself well and enjoy working in a team environment.

In dancing, mechanics and style are two different, important traits. The mechanics of dance include objective, precise execution of the proper footwork and elastic hand contact. For example, mechanically, Salsa and

Mambo have the same cadence of 'quick, quick, slow,' but different timing sets them apart. Stylistically, both dances include stylistic flairs such as moving the arms, extending the fingers, turning the head, and pointing the toes. It's possible—and common—to execute footwork and handwork correctly (mechanics) but look awkward to others and feel awkward to your dance partner (poor style).

The opposite is possible—looking competent from a distance (style) but with handwork or footwork that fails to match the music (mechanics). Good mechanics without style is robotic and boring, while style with poor mechanics resembles chaos, not partnering.

A responsible dancer focuses on mechanics until those feel natural, then turns their attention to style. A dancer who emphasizes style over mechanics is aloof and disrespectful. A dancer who emphasizes mechanics first and then develops their sense of style shows healthy self-management.

Mechanics and style are distinct in innovation teamwork, too. Mechanics pertains to the *objective* execution of the three communication channels (meetings, emails, and asset documentation). Among the channels, healthy self-management is *punctuality* to meetings, *responsiveness* to email, and the *fulfillment* of an assignment (i.e., one of Five Verbs) for an asset.

Mechanics also concerns the asset inventory. Good mechanics require completing the right assets, synchronized in the right sequence, at a sustainable pace. Any work beyond these communication channels and asset inventory is waste; it dilutes attention from durable work output. Innovation teams and individuals must perform these mechanics well before they turn their attention to elements of style.

Style pertains to the *subjective* execution of the three communication channels. An employee's meeting style includes body language such as engagement and eye contact. An employee's email style relates to their length and tone. In contributing to assets, a stakeholder might gravitate toward one of Five Verbs (e.g., drafting or approving). Or an employee might gravitate toward one pool of assets (e.g., people or process).

An innovation professional's style is often seen when they apply their past experience to teamwork. Someone might apply expertise from different industries (e.g., healthcare or agriculture), whereas someone else might apply expertise from another business function (e.g., marketing or customer service). Style in innovation often comes from cross-pollination with an unrelated experience.

Retention—particularly of learned mechanics—matters in dance and in innovation. It's frustrating when someone forgets what they learned from one dance rehearsal to the next. Although frustrating, it's common. Instructors routinely re-teach things and remind students to practice material outside of class to maximize retention.

Innovators get frustrated with poor retention, too. Rehashing topics and decisions hurt everyone's progress. Maximize retention—for the benefit of yourself and your future team—by writing things down. Ideally, dancers and innovators minimize reteaching and rehashing topics; that way, every session can build upon the last.

For dance and innovation work, mechanics are *what* you bring, and style is *how* you bring it. Style makes the dance and innovation experience more unique because it is an unexpected source of passion, learning, and entertainment. Style fuels positive surprises that also improve the customer experience. Ideally, innovation professionals master mechanics, then go beyond 'just going through the motions' and infuse their style.

Mechanics, style, and retention of what you learned comprise healthy self-management. These emulate your actions within an innovation team and are a foundation for other aspects of teamwork.

Self-Awareness

Opportunity dances with those already on the dance floor.
~ H. Jackson Brown, Jr. (1940-2021), author of *Life's Little Instruction Book*

Skilled dancers and innovation professionals manage their bodies and what's inside their heads. You should also have a realistic self-awareness of your goals, fears, strengths, and weaknesses.

Dancers should be thoughtful about why they dance. The reasons to take up dance might be superficial or profound. One simple reason is to escape cabin fever. Profound reasons include quality time with friends or a significant other. Any reason at all is great, but having *no* reason diminishes you and your partner's experience. If others don't know what you're all about, they won't explore with you where you fit best, and your journey muddles along aimlessly. When others know your reason—your *raison d'être*—for dance, they help you get there, and you accelerate your investment in the art.

Innovators must also be thoughtful about why they innovate. Simplistic reasons like the need for a job are common and perfectly fine. Profound reasons, such as positively impacting the lives of your customers and stakeholders, are also great. But it's essential to know your innovation *why*.

For many people, dancing is terrifying. Many beginners only have exposure to advanced dancers and lose sight of the fact that everyone was at some point a clumsy and terrified beginner. Instructors and senior dancers are responsible for welcoming beginners and giving them the confidence so that they will soon be glad to be there. Beginners learn better when fear doesn't cloud their heads.

Innovation work can be terrifying as well. A new team member might only be exposed to the most advanced topics and buzzwords, and forget that everyone had to learn business and innovation basics. Good leaders and empathetic team members welcome beginners and give them the confidence that they will soon be comfortable in their role. Beginners contribute faster when fear doesn't cloud their heads.

Things are often intense for dancers and innovators while in beginner status. Being a beginner on the dance floor is exciting, but there's no incentive to extend your time as one. It's natural to want to advance quickly. Taking lessons as intensely as your schedule and other obligations allow can reduce the window of being a beginner. Beginners on an innovation team often 'drink from the firehose' and intensely learn their new role. The intensity isn't a bad thing; it's a temporary endeavor that gets them up to full strength more quickly than casual exposure to the work.

A flashy dancer might impress the novice dancer, but flashiness does not automatically translate to a good fit on the dance floor. Some advanced dancers do not enjoy dancing with beginners. It's okay to admire them from a distance but minimize your attention on them. Flashy dancers are self-centric, so they rarely make great partners.

An innovation professional's pedigree and bluster might impress, but they don't guarantee a good fit as a leader or teacher. The best leaders and teachers minimize intimidation and attention on themselves. They maximize confidence and enthusiasm in their followers, on the mission, and in the collaboration.

Humility is an attractive trait on the dance floor. Once they get a taste of progress, some dancers' egos outpace their talent, earning them the label of poser. Posers repel others. In a group setting, always assume someone is more talented than you in some aspect of dance. Be an admired dance partner in others' eyes, not your own.

Humility is also an attractive trait in an innovation team. Humility allows a team to focus on work without distraction from a poser's demeanor. Posers garner insincere flattery, but genuine humility opens the door for sincere compliments, positive reinforcement, and additional invitations to collaborate.

Being a Good Leader

A leader is simply a follower who follows what they believe in.

~ Unknown

A fundamental, unforgiving truth about partner dance is that the art requires a leader and a follower. Leaders and followers have different responsibilities. And since dancers rotate partners, leading and following feel different among multiple partners. But responsibilities are more balanced than what most expect.

On the dance floor, a leader's first task is to welcome the follower. A smile and eye contact are all that's necessary. A leader's welcome conveys optimism and confidence in themselves and their partner.

In innovation, a leader's first task is to welcome new followers. The welcome is more than body language; it can be a short conversation. Good leaders convey optimism and confidence in themselves and the team.

As a dance couple moves, the leader's next goal is clarity. Clarity is empathy. A leader's clarity shows they care about the follower's experience. Noise in the leader's motions sets the follower up for failure. Clarity makes the follower's job—steps, turns, rotations—easy. Clarity helps a follower to transcend mechanics and pay attention to their style.

A good innovation leader provides clarity. Ambiguity in innovation teamwork sets employees up for failure. Clarity makes employees' mechanics—meetings, email, and asset creation—easy. Clarity helps employees transcend mechanics, apply their expertise and experiences, and inject their unique style and creativity. Clarity sets employees up for success.

Spatial awareness and boundaries are fundamental in dance. Poor boundaries include invading someone's space, stepping on someone, or injuring them. Leaders on the dance floor are aware and respectful of personal space. Leaders are responsible for keeping their followers and themselves safe and not impeding other couples from dancing.

Innovation leaders must likewise keep team members in their appropriate space to avoid collisions and injury. Assignments, schedules, and status reports ensure team members maintain boundaries and don't hinder each other from doing their job.

On the dance floor, the leader is accountable for mistakes and recovery. When a leader realizes they have made a mistake, the gracious reaction is to acknowledge it, reset, and move on. If a follower misses a cue, the leader instinctively graces past it and contemplates what they could have done differently so that the follower won't miss the cue again.

A common adjustment a leader makes is to manage mechanics with a stronger hand. For example, if a follower misses the cue for a turn, the leader should exaggerate the signal to turn and minimize other noise they cause.

Innovation leaders are accountable for mistakes and recovery. When good leaders realize they have made a mistake, they lead by example by acknowledging it and moving on. When a team member misses a cue, a gracious leader considers what they can change to minimize the chance that a team member will miss the cue again. Common adjustments include increasing detail in project plans, status reports, and performance feedback exercises.

This resembles a dance leader managing mechanics with a stronger hand. For example, if a team member misses a cue, a strong leader exaggerates the signals to contribute (Five Verbs) and doesn't tolerate noise (verb sprawl).

Small mistakes on the dance floor resemble sampling errors. Small mistakes are often a source of learning for the leader and the follower. Big mistakes are systemic bad habits in a dancer's mechanics. The most common bad habits are too strong of a hand-hold, too weak of a hand-hold, and poor rhythm in footwork.

Small sampling errors in innovation are sources of learning for leaders and followers. Bigger, systemic, mistakes in innovation also typically pertain to mechanics—a team's emphasis among the three communication channels and selection of assets. Teamwork should continue when everyone can resume with what the leader has in mind. The leader fixes systemic errors by changing the assignments of Five Verbs.

A good dance leader makes the most of every dance. They meet their follower where they are by accommodating the follower's skill level. For the duration of the dance, the leader tries to be who the follower needs them to be. A poor leader disregards their partner's skill level.

An innovation leader has a similar relationship. They bring out the best in their followers by meeting them where they are according to their capabilities. They adjust to the follower's skills and growth potential, while a poor leader disregards an employee's expertise.

Within a few seconds of dancing, the leader can typically tell whether they should lead with a heavy hand or a delicate touch. If the follower signals they need a heavy hand, the leader emphasizes mechanics. If a follower signals that a light touch is sufficient, both can emphasize style. The effortlessness shows high competence to an audience. Someone's favorite dancer is often determined by such high compatibility that mechanics need minimal brainpower, and instead both dancers gracefully style their way through a song.

A mature innovation leader is a minimalist too—managing just enough to keep risk modest, execute sound mechanics, and encourage each team member to inject their own flair. Mature innovation leaders make collaboration look effortless.

And finally, a good leader dances with authenticity. They know whether or not a dance is going well. A pleasant dance contains mutual approval; even for an awkward dance, an empathetic leader still shares a sincere thank you. Pleasant or awkward, authenticity is better than insincerity.

A good innovation leader also exhibits authenticity. They know whether or not a project is going well. Team members can sense project health, a leader's authenticity, or a leader's cluelessness. A good innovation leader knows what success looks like, acknowledges when collaboration feels awkward, and asks themselves what they can do differently.

A dance leader's checklist is also an excellent list for every innovation leader. Lisa La Boriqua, Chicago's Latin Street Dance Studio co-founder, explains in every dance lesson, "If your partner is smiling, you're doing it right!"

Being a Good Follower

I am reminded how hollow the label of leadership sometimes is and how heroic followership can be.

~ Warren Bennis (1925-2014)

Someone unfamiliar with partner dance might suppose followers on the dance floor have it easy. More often than not, that's false. As the American cartoonist Bob Thaves pointed out about Fred Astaire, "Ginger Rogers did everything *he* did ... backwards and in high heels."[15]

Often, innovation followers don't have it easy, either. When they feel like they're working backward in high heels, followers would benefit from taking a cue from their counterparts on the dance floor.

Just like leaders need to execute mechanics and style, followers do, too. Followers can encounter leaders with messy mechanics and style. Skilled followers resist emulating them and keep performing with integrity and elegance. A skilled leader makes mechanics easy for followers, enabling them to pay attention to style. A skilled innovation follower can execute their mechanics regardless of their leader's skills.

A dance follower returns the favor of making the leader's job easy by having a frame. A follower creates a frame by shaping their hands, arms, shoulders, and torso so the leader can hold and navigate them around the dance floor. A frame is the follower's personal dance space. The ideal frame is elastic and responsive—never stiff or floppy.

Innovation followers also create a frame by reporting status and workload level to their leader. These convey an employee's boundaries for additional work. The ideal frame for a follower is never stiff or floppy, but firm and elastic. A skilled innovation leader elegantly weaves the talents of followers around a project to exercise and build their skills. A firm, elastic frame makes the leader's job easy.

A good dance follower is active and responsive to the leader's cues. The leader makes decisions for two people every few seconds. The return pressure in the follower's hands and arms instructs the leader whether to lead with a light touch or with a heavy hand. The follower's facial expressions tell the leader how well they're leading, how well the follower can execute,

---

15 Ginger Rogers and Fred Astaire were a celebrity ballroom dance couple in early twentieth-century Hollywood movies.

and if the follower is enjoying the dance. Continual responsiveness shares crucial information with the leader.

Active followers are also necessary for a healthy innovation team. An adept innovation leader welcomes cues from their followers. They signal to their leaders whether they need a heavy hand (micromanaging) or a delicate touch (guided autonomy). A follower's verbal and body language tell the leader how well they lead and if the work is enjoyable. Passive team members hurt speed, shrug at their own ideas, and ignore feedback they have for leaders. The best followers are active, responsive to their leaders, and generous with information. On and off the dance floor, active followers are more valuable.

It's rare, but sometimes a dance leader fails because of their own incompetence or malevolence. A poser doesn't know the dance, or they mistreat the follower. Accepting this isn't empathy; it's a lack of self-respect. As soon as the follower feels safe, they should abort the dance. Followers in innovation should exercise the same self-respect. If a leader is incompetent or abusive, the follower should feel licensed to abort their role.

The best followers are ready for variability among their dance leaders. Early in a dance, a follower can gauge whether the leader is easy to follow and calibrate their attention across mechanics and style. Complex dances cause the follower to pay attention to mechanics. Simple dances allow the follower to get creative in styling.

Early in a project, innovation followers can gauge whether their leader is easy to follow. Leaders with low clarity are challenging, forcing followers to pay attention to mechanics. Leaders with clarity are easy to follow and encourage followers to style and get creative.

A good follower exhibits poise on the dance floor. They know that surprises and mistakes happen. Like a good leader, a good follower acknowledges their role in the mistake, adjusts, and resets calmly. Although leaders are responsible for safety, they cannot see everything. Followers are in a great position to notice danger the leader doesn't catch (such as another dance couple about to collide with them).

A good innovation follower likewise exhibits poise in their teamwork. When surprises or mistakes occur, they keep their cool, acknowledge their role in what happened, and get ready to resume with adjustments. Leaders are responsible for safety, but they have blind spots. A good leader welcomes being educated by followers, who are well-positioned to notice blind spots such as a delay, missing assignment, or hidden dependency.

A good dance follower knows the leader also has a tough job. A gracious follower conveys their appreciation as appropriate. The same is true for a good innovation follower. Leaders certainly welcome positive reinforcement from their team.

Finally, a good follower makes the most of every dance. Expectations can soar or plummet during a dance. A follower savors a great dance because it lasts around four minutes. A follower can take consolation during an unpleasant dance because it only lasts around four minutes. Learn what you can to apply another day. A good innovation follower makes the most out of every project, team, and leader. Good or bad, it's not forever. Learn what you can to apply another day.

A dance follower has a long checklist, from mechanics to appreciation. Though far from easy, it was an excellent list for Ginger Rogers and is a great list for innovation followers.

Being a Good Steward

Every good citizen adds to the strength of a nation.

~ Gordon Hinckley (1910-2008), American religious leader and author

As dancers mature in their community, they fret less about their dance partners or themselves, and instead invest in their community. They want to help others to mature from dance beginners to experts. Stewardship promotes culture traits such as inclusivity, interchangeability, and learning. Good stewards keep things moving and sustainable.

In dance, interchangeability takes the form of rotating partners, which is invaluable. If you're not rotating on the dance floor, you are trapped. You cannot explore dancing on your own terms. Neither can your dance partner since they are trapped with you. Other dancers might not even know you're available. By rotating partners, you grow more comfortable dancing with diverse talent levels and styles. Rotation expands everyone's opportunities. Rotation builds robust individuals and a robust dance community. Rotation brings fresh energy to every dance.

Rotation in innovation is also valuable. Becoming trapped in one role makes you vulnerable when the organization inevitably experiences disruption. Conversely, when you are in a single position for a long time, the organization is likely vulnerable to you since you have probably become a single point of failure. Pursuing safety in a single role ironically undermines *everyone's* safety, including yours.

By rotating positions, you increase your ability to serve different teams and customers. You expand your empathy for the various skills needed across an entire project. Rotation encourages you to learn and become more valuable to your future stakeholders.

In partner dancing, you rarely stay—or get stuck—with one partner for long. If you have ten dance partners in a single evening, you have a favorite, a least favorite, and everyone in between. Etiquette encourages you to accept one dance from anyone who asks, even your least favorite. Etiquette also asks the least favorite to be perceptive enough not to invite you to dance again. And when you are sitting one out—setting a boundary for your time—it is always okay. Everyone needs rest, for themselves and their next dance partner.

Healthy innovation professionals prioritize rotation and rest. Innovation professionals should have a low bar in the spirit of 'accepting a single dance.' Most innovation professionals can add value to projects even if they limit their contribution to hours or a few days. On the other hand, they should avoid getting stuck with an organization that's not their favorite. No one is obliged to remain with a bad innovation partner or in a toxic environment. And when innovation professionals need a break, they should be comfortable sitting out an opportunity and decline graciously. Everyone needs rest, for themselves and for their next team.

Considerate dancers know that others are constantly moving around them. Dancers maintain their boundaries, but don't get in the way as others navigate around the dance floor. Respectful dance partners occupy space on the floor for two people and not more. Respectful innovation professionals also maintain a modest 'footprint' around others. Maintain boundaries for yourself, respect others' boundaries, and don't hinder others from navigating their own jobs and careers.

In dance, learning never stops. A beginner dancer moves on to intermediate and advanced lessons. When a dancer plateaus in Salsa, they learn the Cha-Cha. When they plateau in Cha-Cha, they learn the Bachata. They could also move on to Ballroom dances after exhausting all the Latin ones. Dancers continually learn new mechanics and styles.

Innovators also never stop learning. They learn different tools and technologies. They learn about different businesses and industries. Colleagues learn directly from each other. In fact, you are likely an expert at something that a colleague is a beginner in, and vice versa. Learning innovation mechanics and styles is a never-ending journey that holds more fun and fascination than frustration.

The dance culture values courtesy and etiquette. Hostility and rudeness don't survive long in the dance community. Studios and nightclubs demand courtesy to retain membership. A steady rhythm of songs, dances, and partners resembles a graceful factory.

Most innovators don't have to accept toxic colleagues or environments. It is increasingly difficult for rude people—uniquely unskilled and uninterested in empathy and the arts—to survive in the innovation community. Rude workers inspire books like *The No Asshole Rule*. An empathetic rhythm of assignments, projects, and relationships resembles a graceful factory.

Stewards of the dance community monitor and invest in the evolution and growth of a few different forums. These forums include lessons, choreography, and nightclub events. Stewards monitor for circumstances that threaten cultural health and community adaptability. They are receptive to ideas to change and grow.

Experienced innovators also become less concerned about isolated events and care more about the cultures of their communities. Good innovation stewards make team culture compelling, valuable, sustainable, and fun.

Dance provides culture traits that bring out the best collaboration in people. The Elegance methodology emulates the culture of partner dance to realize the benefits in innovation teamwork. The innovation community can learn from these numerous aspects of partner dance culture: how to manage yourself, your partner, and how to care about a community. Everything you learn on the dance floor applies off the dance floor, particularly to teamwork. The elegance in your mechanics and style wins customers. If you're smiling, you're doing it right.

Theater

People may have snickered when we started.
But they'll take photos when we're done.

~ Unknown

Although the software-centric innovation methodologies have been in use for two decades, definitions and usage are far from standardized. Fragmentation and variability are rampant. Innovation leaders reinvent the wheel at the mercy of personalities, politics, and pet projects. Repeated reversals to methodology indicate that an organization is fixing the wrong things.

Because conventional methodologies emphasize technology and data, they demote (or give only lip service to) the human story. People-centric innovation is precisely about the human story—it leverages technology and data in *service* to it.

Integrating people, process, and technology requires abundant creativity. The nature of this integration should always be the same, but instead, innovation teams repeat the same mistakes. There is a pattern to the problems, and the art of theater is a powerful model to stop that pattern.

First, in theater, story is everything. Even exceptional talent, costumes, and props cannot rescue an uninspiring story. Once the story is stable, theater provides a forum to integrate the other elements that build the audience experience. And finally, theater productions at all skill levels execute the same methodology to ensure success. Reinventing the wheel is counterproductive and can lead to failure.

Applying theater to innovation leverages passion and purpose. It strives for an improved story that puts the customer at the center as a main character and hero. Innovation teams don't have to execute methodology to perfection. Good-faith collaboration composes a compelling plot for stakeholders, builds empathy for them, and hits the climaxes and moments that matter. Modeling theater engages employees and customers. Theater is the model that brings the best stories to real life. As Augusto Boal, Brazilian theater director, once noted, "Theater is a form of knowledge; it should and can also be a means of transforming society. Theater can help us build our future rather than just waiting for it."

Story

Innovation is rewriting our customer's story.

~ Unattributed

Empathy is seeing someone's situation and understanding why they want to improve it. Empathy in innovation improves a person's situation. Improving their experiences is effectively rewriting their story. This is why storytelling is such a powerful tool for innovation.

Some storytelling feels like fluff. Theater is a tool to convert make-believe into reality and bring storytelling to life. The foundations for great theater are the story and the script. The story explains how characters evolve, and the script clarifies what the characters say and do.

Some innovations can feel like fluff. Customers are delighted when your innovation team converts make-believe into reality. To accomplish this, empathize with your customer's story, grasp their current script, and write their improved script. Being part of a story shows customers how and where they are relevant.

Theater is a great model to stay focused on your customer instead of yourself. Put your customers at the center of their own story. Write about them and consider writing *with* them.

The story—the actor's context—is the primary focus in theater. The contents of the stage are a secondary focus. In innovation, the story—the customer's context—is the primary focus. Objects like technology, data, and content are a secondary focus.

Theater is an environment that benefits when it has both big-picture thinking and attention to detail. Audiences love the long arc of the story and the numerous mini climaxes among scenes. Innovation benefits in the same way. Customers love the long arc and thoughtful detail in the experiences innovators build for them. Innovation teams can place members in roles that leverage their affinity for detail.

A final parallel between theater and innovation in rewriting stories is that trying to change too much simultaneously is risky. The best rewriting starts small and at the beginning. It avoids changing too much too fast, sees the cascade effect of changes early in the story, and minimizes duplicate work for changes at the end of the story. Innovation teams must tackle changes in modest portions to minimize risk and rework. Partitioning work into modest amounts has the formal term Release Management.

Contributing to rewriting your customer's story is a great model to ensure you focus on their experience at all levels in low-risk, sustainable stages.

Integration

Integration happens when all parts of your being are in harmony.

~ Amy Leigh Mercree (b. 1977), American author and holistic health expert

A great story is necessary for a successful production, but the story itself doesn't guarantee success. A theater production has many parts that make it successful, and the relationships among those parts require integration.

Similarly, compelling, high-priority improvements to the customer story are necessary for successful innovation, but the vision alone doesn't guarantee success. A project has numerous parts that make it successful, and its relationships require integration.

Parts of a theater production that drive its success include choreography, props, costumes, publicity, and stage management. All the work has a mindful sequence and dependencies. Early work is isolated; later work is heavily integrated, especially approaching opening night. Parts of an innovation project include process definition, programming, training, reporting, and coaching. Early testing is isolated; later testing is heavily integrated, especially approaching a go-live event. Integration matters because when things don't work across functions, setbacks in quality and speed occur.

The numerous functions of theater and innovation work require mindful assignments and aggressive decentralization of the work. Casual decentralization puts too much responsibility on too few people, risking burnout and delays. Aggressive decentralization solves that risk but creates a different hazard with communication between functions. Aggressive integration mitigates risk between people handing off information.

Feedback among theater performers is unusually transparent. A typical feedback session is thirty minutes long after a two-hour rehearsal. Everyone sits down to listen to the director, who reads their list of notes aloud. A director's notes contain positive and negative surprises, and everyone hears everyone else's feedback. This open forum breeds high vulnerability, humility, and integration. While the script micromanages each actor, each actor must be approachable for this micro-mentoring. Everyone learns from everyone else's feedback in this efficient event. This notes session is an incredible model for an innovation team and a powerful habit to establish collective accountability across an innovation team.

The culture of integration in theater creates a fantastic sense of belonging for cast members. Cliques are difficult to develop because of such high interdependency among the performers. All cast members passed the audition, so they're pre-approved. The cool kids are everyone, and the entire cast is the clique. Theater is a model for inclusion.

Innovation is also cool; everyone has passed their audition, and the interdependency is high. A sense of belonging is instrumental for innovation teams to achieve high morale and project success rates.

Integration is a vital culture trait for an innovation team. Like a theater production, innovation requires a ton of interdependent work. Decentralize the workload among many people and carefully manage handoffs. Everyone can learn from the feedback everyone else receives. Like theater, innovation is at its best when everyone contributes to a sense of belonging.

Reinventing The Wheel

If I have seen further than others, it is by standing on the shoulders of giants.
~ Sir Isaac Newton (1643-1727), English scientist and mathematician

The act of putting on a play has an unusually long precedent. Humans have done it for millennia. The process is more complicated than in centuries past, but every theater production learns from its predecessors. For a methodology to put on a play, there is no reason to reinvent the wheel.

Innovation work also has an unusually long precedent. Humans have been doing it for a long time, including a dizzying amount in just the past hundred years. Conventional methodologies aim to minimize reinventing the wheel, but innovation leaders have yet to leverage past mistakes, learn enough from their predecessors, and fix the right things. Problems persist, and they have patterns. Common conclusions include, "We built what the customer asked for but not what they needed," and "We burned the team out trying to meet the deadline." The never-ending excuses include, "We didn't communicate well," and "Things are always more complicated than you plan for."

Good innovation leaders emulate theater directors and *don't* reinvent the wheel. The Elegance methodology aggressively anticipates the patterns and fixes the right things, so projects minimize mistakes of their predecessors and are always more successful.

Every theatre director starts out knowing all the work that every play needs. Likewise, the Elegance methodology lays out the work every innovation leader needs. Cutting corners in either profession is disastrous and sets up ensembles to fail.

Preventing the need to reinvent the wheel requires the right wheel in the first place. Having the correct wheel in place is valuable because it lowers marginal cost. Reinventing the wheel makes projects significantly more expensive than they should be. Avoiding it liberates attention from the mechanics of project management to pay more attention to customers instead. Junior employees can take over administration from senior, more expensive team members. This attention shift is entrepreneurial as senior employees chase stakeholder empathy, value creation, and revenue streams.

For example, between New Year's Day and Spring Prom, high schools in the United States that put on a play execute the same annual schedule. This schedule is fourteen weeks long. The schedule has high visibility, high stability, and low ambiguity. At the start of this schedule, everyone knows their role and dependencies. Any other structure sets the ensemble up for failure. Embracing this schedule frees them from reinventing the schedule. Elegant innovation teams imitate this. A high success rate requires a schedule with high stability and visibility and low ambiguity and duration.

This freedom is a form of automation. It resolves systemic and systematic errors, increasing success rates. It gives grace to sampling errors, increasing empathy. And it gives grace to entrepreneurial professionals just trying to author a better story for their customers.

Great theater suspends your sense of reality, but every play *does* end, and the audience goes home and returns to real life. Theater is a simulation; the artists and audience feel euphoria and pain without permanent consequences.

Innovation benefits from the same simulation. Simulation occurs in projects in the testing phases. Approach testing, simulation, and innovation as theater and you will uncover the compelling changes to a story. Building moments that matter for stakeholders requires you to aggressively rewrite your customers' stories and integrate the moving parts without reinventing the wheel. These steps transform your innovation efforts into a well-executed play.

At its best, your innovation team resembles a theater company. It is story-driven, not data-driven. It has empathy for the characters—people—and an entrepreneurial spirit to improve their story. Integration of the moving parts can confidently proceed because it is automatic, stable, and visible.

The Elegance methodology accounts for all the typical mistakes of past projects. It minimizes the laboriousness of project management so you can collaborate with your customers at your frontier of innovation.

CONCLUSION

Do we fight for the right to a night at the opera now?

~ Enjolras to his fellow students in the musical
Les Misérables

The empathetic arts provide relatable language, habits, and culture for innovation. Innovation teams already use some terminology from the arts, and they can benefit from adopting more. The empathetic arts remind us that our work is not about us.

The arts combat VUCA: the culture of Aikido neutralizes volatility, and the culture of improv embraces uncertainty. The cultures of parenting and theater accommodate complexity, and the cultures of music and dance tame ambiguity.

The metaphor of a factory also combats VUCA. Although your team is not a conventional factory, you should manage it like one because you care about all the same things: speed, quality, waste, and the other topics discussed in this book. Discouraging verb sprawl and formally managing via Five Verbs provides uncommon and ruthless discipline. Emphasis on documentation fosters clarity, simplicity, and transparency for your team today and tomorrow. The asset portfolio lowers ongoing communication costs and improves the economics of your agreement factory.

Teams benefit from adopting habits from both cultures, executing a graceful factory, and building an asset portfolio. The portfolio of the arts is evidence of a culture that cares about its customers and employees. Tools to conquer VUCA build confidence and courage in innovation teams.

The metaphors in the Elegance methodology are disguises for your innovation team's culture. You now have eight disguises: a factory, the asset portfolio, and the six arts. These disguises support the 'why' behind your project management culture. For example, your support of employee self-sufficiency says, "This is culture disguised as a parent." Your support of reliable work rhythm is, "This is culture disguised as a factory."

Conventional innovation methodologies have high failure rates and tolerate poor employee experiences. They emphasize working software, which is not what makes innovation difficult. What makes innovation difficult is the need for a working team. The Elegance methodology emphasizes working teams. Like an ensemble in the empathetic arts, working teams have a great employee experience and high success rates.

The main ingredient for a working team is mindfulness for when team members should compete and when they collaborate. When not mindful, team members compete at the wrong time about the wrong things. They make the same mistakes with collaboration. A healthy team culture 'sets the table' and 'bangs the table' for the right competition and collaboration times. Accomplishing this is both a competitive advantage and a collaborative advantage.

You are now armed with two of the three metaphors to create a culture of Innovation Elegance. The metaphors of a factory and the arts explain the culture traits your team needs to collaborate with ruthless discipline and graceful empathy.

The third metaphor—an asset portfolio—is thoroughly explored in *Innovation Portfolio*, the second volume of *Innovation Elegance*. *Innovation Portfolio* prescribes documentation so durable and valuable that each item warrants the term 'asset.' Each asset reinforces a culture of discipline and empathy. Each asset is *culture disguised as a template.*

The term ruthless grace sounds harsh, abstract, and contradictory. But consider: the culture traits of a factory are unusually disciplined—arguably ruthless. Culture traits in the arts are exceptionally forgiving and typically graceful. Although ruthlessness and grace seem to be opposites, they are complementary and contain much of the same language. Teams, businesses, and society demand that innovation leaders have complementary skills and tools to be aggressive with process and gentle with people.

Any innovation leader using a software-centric methodology such as Agile is in debt—a *methodology debt.* To get out of debt, innovation leaders must transcend Agile to adopt a methodology that manages what makes innovation difficult: people.

Innovation teams with no methodology debt are at a frontier—a frontier to innovate with freedom. The Elegance methodology rethinks how we interact, govern, and collaborate.

Many innovation leaders already know the value of discipline and empathy in their culture. But their tools tolerate ambiguity as they attempt to orchestrate ruthless grace. The Elegance methodology provides tools and clarity for innovation leaders to operate with courage, confidence, and career security.

AFTERWORD

This book emphasizes culture, defines culture as shared behavior, and explains the Five Verbs framework at the cultural level. This book did not get prescriptive about where to apply Five Verbs and how it supports the metaphor of the asset portfolio. That is the focus of a separate and complementary book—*Innovation Portfolio*.

A new innovation methodology is disruptive. It challenges businesses to innovate how they innovate. As the Elegance methodology reduces the marginal cost of innovation, innovation professionals will be motivated to work earlier in the innovation lifecycle—closer to customers—creating and improving customer experiences.

Like many innovations, Elegance reduces work for some jobs and increases work for others. Like advanced dance partners, innovation teams will spend less time on mechanics and more on style. Teams will have discipline in their muscle memories, saving energy for their chosen style of empathy.

As you probably know, many people look at the reviews on Amazon before they decide to purchase a book. If you liked the book, please leave a review with your feedback. It takes only a few minutes, and it helps future readers.

I want to know in what ways this book was helpful to you and your team. Please find me on LinkedIn at www.linkedin.com/company/innovation-elegance/ where you can learn more how I apply Innovation Elegance to all sorts of ensembles.

Thank you very much,
Robert Snyder

Author Biography

Debut author Robert Snyder is a thirty-year project management and performing arts veteran who aims to make healthy change straight-forward and innovation success inevitable. He hears the dissatisfaction with poor employee experience and unacceptable project failure rates and knows first-hand the limitations of software-centric methodologies.

Robert believes innovation professionals are ready for a fresh, people-centric methodology that confidently administers unusually rigorous discipline *and* graceful empathy. Robert's career path includes consulting and corporate roles, PMP and Agile certifications, and countless performances in vocal, dance, and theater ensembles. Robert earned his BS in Electrical Engineering from the University of Illinois and his MBA in Strategy and Analytical Consulting from the Kellogg School of Management at Northwestern University.